AF183342

PSYCHOLOGY RESEARCH PROGRESS

SPATIAL ORIENTATION IN HUMANS AND ANIMALS

Psychology Research Progress

Additional books in this series can be found on Nova's website
under the Series tab.

Additional e-books in this series can be found on Nova's website
under the e-book tab.

PSYCHOLOGY RESEARCH PROGRESS

SPATIAL ORIENTATION IN HUMANS AND ANIMALS

ROBERT LALONDE, PH.D.
EDITOR

nova publishers
New York

Copyright © 2014 by Nova Science Publishers, Inc.

All rights reserved. No part of this book may be reproduced, stored in a retrieval system or transmitted in any form or by any means: electronic, electrostatic, magnetic, tape, mechanical photocopying, recording or otherwise without the written permission of the Publisher.

For permission to use material from this book please contact us:
Telephone 631-231-7269; Fax 631-231-8175
Web Site: http://www.novapublishers.com

NOTICE TO THE READER

The Publisher has taken reasonable care in the preparation of this book, but makes no expressed or implied warranty of any kind and assumes no responsibility for any errors or omissions. No liability is assumed for incidental or consequential damages in connection with or arising out of information contained in this book. The Publisher shall not be liable for any special, consequential, or exemplary damages resulting, in whole or in part, from the readers' use of, or reliance upon, this material. Any parts of this book based on government reports are so indicated and copyright is claimed for those parts to the extent applicable to compilations of such works.

Independent verification should be sought for any data, advice or recommendations contained in this book. In addition, no responsibility is assumed by the publisher for any injury and/or damage to persons or property arising from any methods, products, instructions, ideas or otherwise contained in this publication.

This publication is designed to provide accurate and authoritative information with regard to the subject matter covered herein. It is sold with the clear understanding that the Publisher is not engaged in rendering legal or any other professional services. If legal or any other expert assistance is required, the services of a competent person should be sought. FROM A DECLARATION OF PARTICIPANTS JOINTLY ADOPTED BY A COMMITTEE OF THE AMERICAN BAR ASSOCIATION AND A COMMITTEE OF PUBLISHERS.

Additional color graphics may be available in the e-book version of this book.

Library of Congress Cataloging-in-Publication Data

ISBN: 978-1-63463-001-6

Published by Nova Science Publishers, Inc. † New York

CONTENTS

PREFACE

The volume entitled "Spatial Orientation in Humans and Animals" edited by Dr. Robert Lalonde, from the department of Psychology at the University of Normandy, France, provides a summary of our current knowledge on the neurobiological mechanisms underlying the way human beings and animals orient themselves in their everyday environment. In chapter 1, Lalonde presents the neuropsychological tests used to study visuospatial processing in human subjects. In chapter 2, Lalonde and Dr. Catherine Strazielle, from the Physiology and Anatomy Section at the University of Lorraine, France, present the behavioral tests used to study visuospatial processing in animals. In chapter 3, Strazielle presents the anatomical underpinnings of visuospatial processing, comprising cortico-subcortical circuits that include the posterior parietal cortex and the hippocampus. In chapter 4, Lalonde describes the spatial deficits found in patients with Alzheimer's disease and in chapter 5, the same author along with Strazielle and Dr. Ken-ichiro Fukuchi, from the department of Medicine at the University of Illinois at Peoria, USA, describe the maze deficits found in animal models of Alzheimer's disease. In chapters 6 and 7, Lalonde enumerates the spatial deficits encountered in patients with basal ganglia disorders, such as Parkinson's disease and Huntington's chorea, together with the animal models used to mimic these conditions. In chapter 8, Lalonde describes the spatial deficits found in patients with disorders of the cerebellum. In chapter 9, the final one, Lalonde and Strazielle describe the maze deficits in animal models of cerebellar atrophy. Altogether, this book provides a unique summary of our present knowledge about the neural basis of spatial orientation both in human subjects and animal models.

In: Spatial Orientation in Humans and Animals
Editor: Robert Lalonde
ISBN: 978-1-63463-001-6
© 2014 Nova Science Publishers, Inc.

Chapter 1

SPATIAL TASKS USED IN PATIENTS WITH BRAIN DISORDERS

Robert Lalonde*
Université de Rouen, Dépt. Psychologie, Mont-Saint-Aignan Cedex France

1. INTRODUCTION

Visuospatial functions may be examined at several levels, including judgment of line orientation, visuoconstructive abilities, spatial reasoning, mental rotation, maze learning, map reading, and memory. For all of these functions, a fundamental role is attributed to the posterior parietal cortex, especially in the right hemisphere (Botez et al., 1985; Cytowic, 1996). The posterior parietal cortex receives spatial information from primary visual cortex via the occipito-parietal pathway (chapter 3), in contrast to the occipito-temporal pathway underlying shape recognition (Mishkin and Ungerleider, 1983; Schneider, 1969).

2. JUDGMENT OF LINE ORIENTATION

The Benton judgment of line orientation is among the most prevalent tests used for examining perceptual functions, whereby subjects are asked to evaluate line angulations (Benton et al., 1983). Patients with damage to the right posterior hemisphere are much more vulnerable in this test than those with damage to the left (Lezak, 1983). In depth-neuroanatomic analyses revealed that most deficits arise from lesions in right posterior parietal cortex (Tranel et al., 2009). In contrast, patients with lesions of the prefrontal cortex (Teuber and Mishkin, 1954), or an afferent region, the posterolateral thalamus (Barra et al., 2010), are only impaired in judging the orientation of an illuminated vertical bar in the dark while their head is tilted laterally, called the Aubert effect.

* Email: robert.lalonde@univ-rouen.fr

3. Visuoconstructive Abilities

Visuoconstructive abilities are evaluated on performance subtests of the Wechsler intelligence scale (IQ), notably *block design* and *object assembly*, two highly intercorrelated tests sensitive to lesions in the right hemisphere (Cohen et al., 1988; Lezak, 1983). In the block design test, the subject manipulates plastic blocks to reproduce two-dimensional designs set on paper. In the object assembly test, the subject constructs a puzzle from cardboard figures. In addition to blocks, designs may be reproduced with sticks (Butters et al., 1972). By asking subjects to construct *stick patterns* (Figure 1) under rotated or non-rotated conditions, neuropsychologists can distinguish allocentric abilities based on extrapersonal space from egocentric abilities relative to the subject's own body. The correct reproduction of stick patterns was impaired following damage to either parietal or frontal association cortex.

In addition to object manipulations, visuoconstructive abilities are examined by asking subjects to draw simple objects such as a house, a clock, or a bicycle. The *Rey complex figure* is one of the most often used drawing test, along with the equivalent Taylor figure (Lezak, 1983; Osterrieth, 1944; Rey, 1965). The accuracy in drawing the Rey complex figure has been quantified on a 36-point scale based on the correct reproduction of 18 figure elements, for which age-related normative data are available.

Visuoconstructive scores are liable to deteriorate following damage to posterior parietal (Botez et al., 1985) or frontal (Lhermitte et al., 1972) association cortex. While patients with damage to posterior parietal cortex have difficulties in apprehending a gestalt of the figure, those with damage to prefrontal cortex have difficulties in step-by-step organization (Lezak, 1983). When planning is provided by the experimenter following step-for-step methodologies, the patients' performances on block design and Rey complex figure tests were much improved (Lhermitte et al., 1972). In addition to the neocortex, subcortical regions are involved in visuoconstructive abilities. In particular, case reports indicate deficient copying of the Rey complex figure after damage to the thalamus (Graff-Radford et al., 1984).

4. Spatial Reasoning

Raven standard progressive matrices and Raven colored progressive matrices are time-honored methods to evaluate spatial reasoning (Raven et al., 1984a,b). In Raven standard progressive matrices, subjects view black and white patterns and must choose the one that fits most logically inside the main figure to provide a homogeneous impression. The Raven colored progressive matrices test is a simplified version based on colored figures. Age-related normative data are available for Raven standard progressive matrices (Lezak, 1983). Raven colored progressive matrices are especially vulnerable to right posterior lesions (Berker and Smith, 1988; Costa, 1976), although right anterior regions may also contribute (Kertesz and Dobrowolski, 1981). Another method of evaluating spatial reasoning includes the mental folding test, in which the subject mentally constructs a three-dimensional figure based on two-dimensional drawings containing unfolded sides which must be folded up (Fehrenbach et al., 1984).

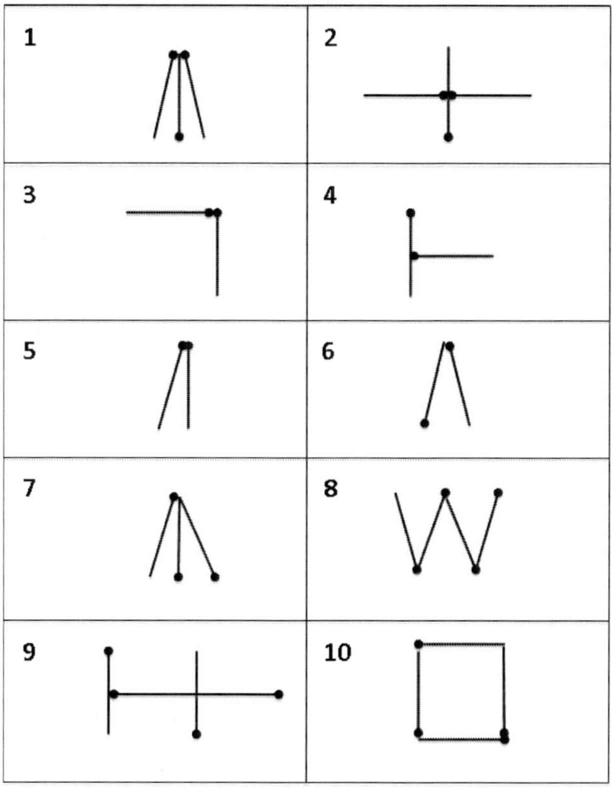

Figure 1. The stick test: the subject manipulates sticks to reproduce a series of geometric designs (Butters et al., 1972).

5. Maze Learning

Visual and tactile mazes have been designed whereby the subject traces by hand a route from start point to goal. One of the better known is the Porteus maze, in which paths must be traced on sheets of paper with the use of a pen (Porteus, 1965). In the stylus maze (Figure 2) (Milner, 1965), the subject uses a stylus to touch metal bolt-heads along a path towards the goal with each error signaled by a noise and the path in plain sight. A tactile version is available with the path hidden from view (Figure 3) (Milner et al., 1968). In addition, a series of covered maze problems have been designed with pathways hidden from sight (Figure 4) (Wallesch et al., 1990). Mazes of different designs have been adapted for use on a touch-screen monitor (Ott et al., 2008).

The acquisition of the stylus maze task is more sensitive to damage on the right than the left hemisphere (de Renzi et al., 1977a). In particular, trials to criterion increased relative to controls after lesions of right parietal, right temporal, or bilateral frontal cortex (Milner, 1965). Both visual and tactile maze problems were impaired in case HM characterized by anterograde amnesia following bilateral lesions of the medial temporal lobe (Milner et al., 1968). In contrast, a patient with damage to temporal neocortex caused by herpes simplex encephalitis had impaired semantic memory but intact stylus maze and Porteus maze performances, presumably due to sparing of the medial temporal lobe (McCarthy et al., 1996).

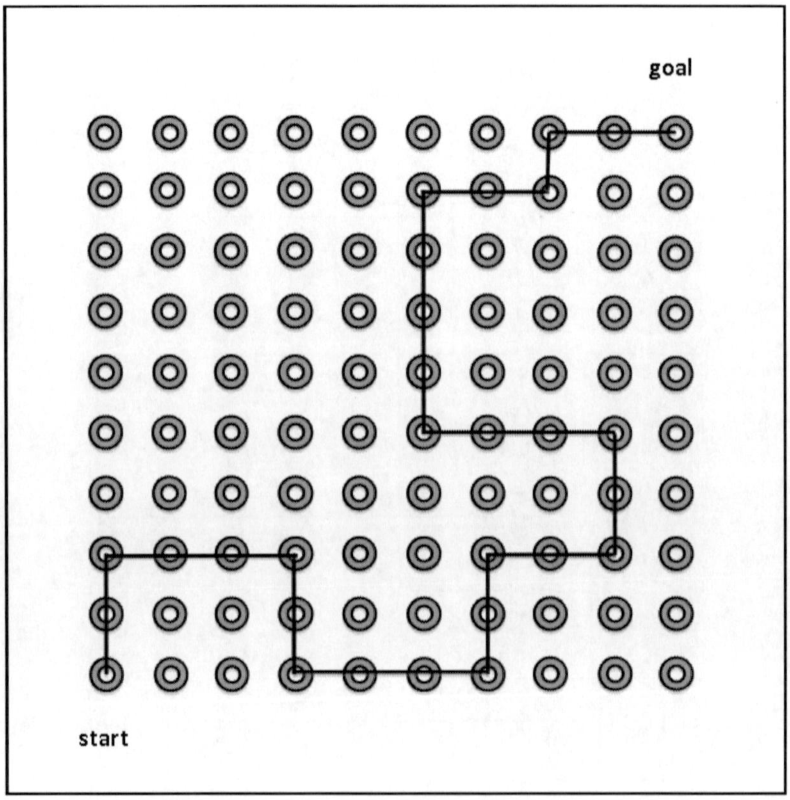

Figure 2. Visual stylus maze: the subject touches metal bolt-heads along a path from the start point to the goal while seeing the path throughout (Milner, 1965).

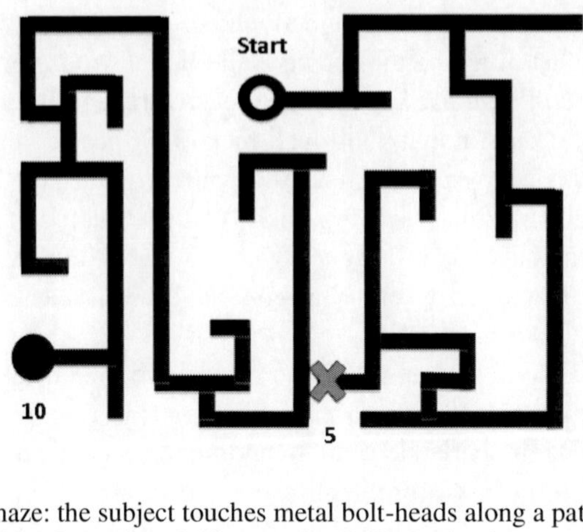

Figure 3. Tactile stylus maze: the subject touches metal bolt-heads along a path from the start point to the goal with the path hidden from view (Milner et al., 1968). In the short version, ending at point X, there are 5 choice-points, in the longer version, 10.

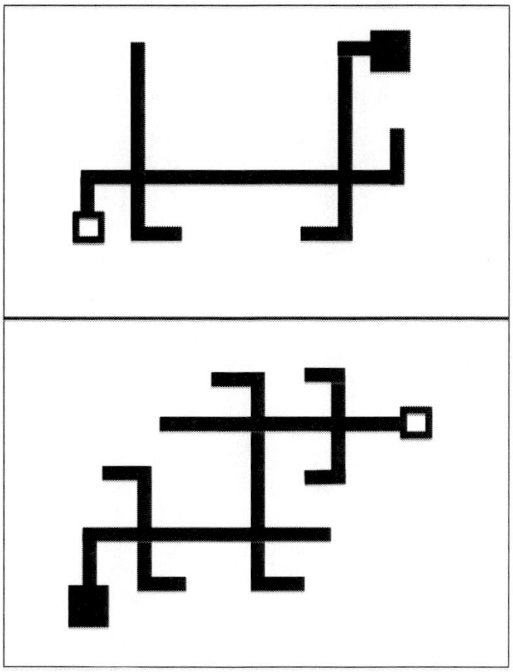

Figure 4. Two examples of covered maze problems with the goal out of sight from start (■) to finish (❑) (Wallesch et al., 1990).

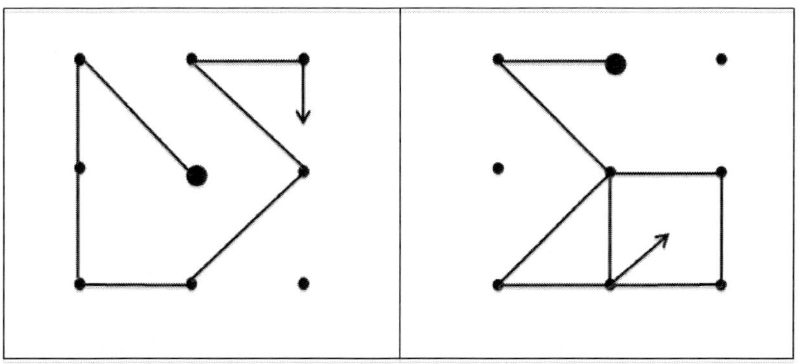

Figure 5. Two route-walking patterns requiring whole body movements (Bowen et al., 1972): the subject is asked to walk along pre-determined paths and finishes where the arrow is pointing.

In addition to maze problems solved by hand movements, those requiring whole body movements have been designed (Figure 5) (Bowen et al., 1972) analogous to those used in animal research (chapter 2). A multi-unit maze for use in human subjects has been constructed (Chouinard et al., 2003), as well as circular arenas akin to the Morris water maze (Hort et al., 2007; Majerová et al., 2012). Moreover, walking along corridors of a hospital ward (Monacelli et al., 2003) and driving a motored vehicle along the streets (Uc et al., 2004) have been used to assess spatial orientation in real-life settings.

6. Map Reading

Different aspects of visuospatial functions are examined by requiring subjects to interpret maps. In the Money road-map test (Figure 6), the subject views two routes taken by hypothetical travellers along streets and intersections (Money, 1976).

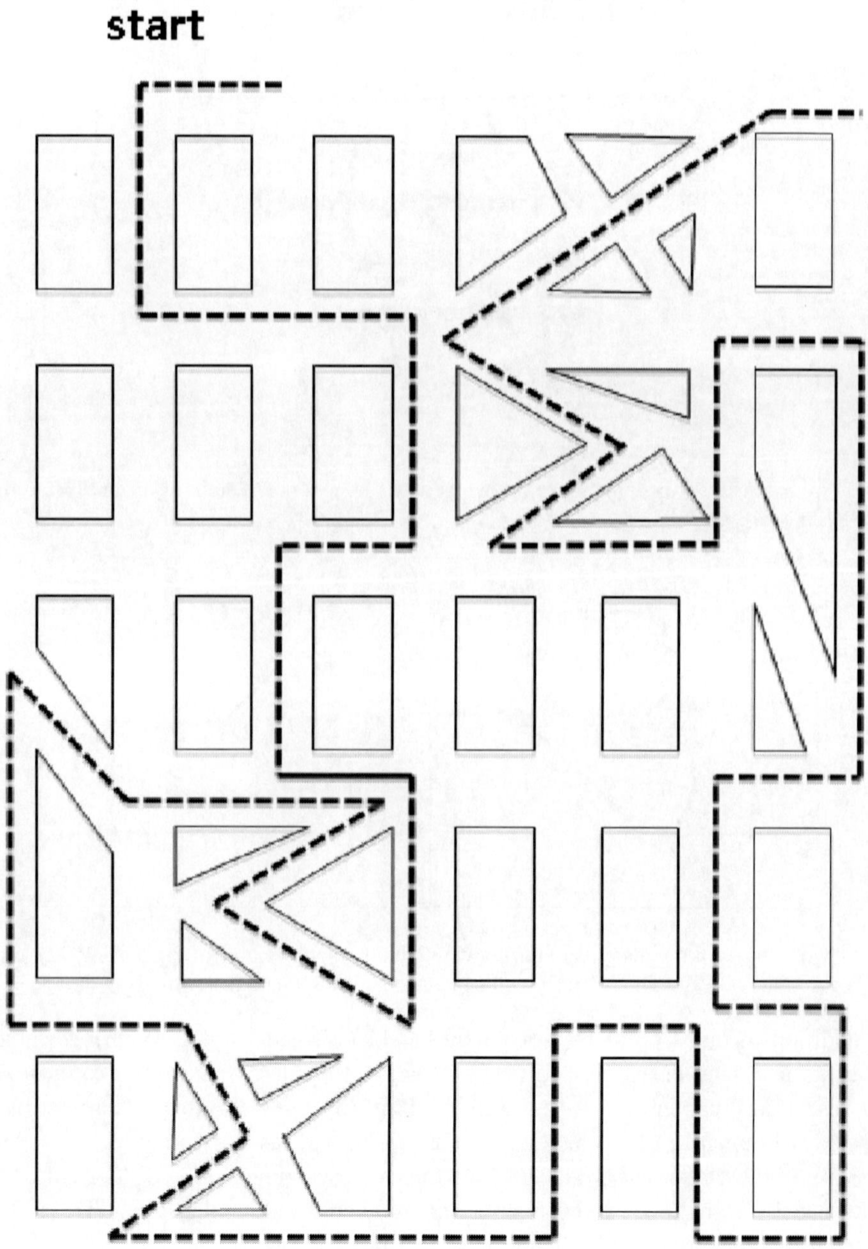

Figure 6. The Money road-map test: the subject views a route taken by a hypothetical traveller along streets and intersections from start to finish (Butters et al., 1972).

At each intersection, they select to turn left or right on their way towards a specific destination. This task is considered to be an egocentric (viewer-dependent) as opposed to an allocentric (viewer-independent) function, because it depends on spatial cues relative to the subject's body. It is sensitive to lesions of parietal and frontal association cortex (Butters et al., 1972). More recently, two route-finding tasks inside a virtual town have been developed: the virtual route learning test and the heading orientation test (Pengas et al., 2010).

The New map test (Beatty and Tröster, 1987) is used to assess retention of novel as opposed to old information. The subject views a map of three hypothetical states containing the names of 15 hypothetical towns and is asked to locate them by name. In contrast, the Fargo map test (Beatty, 1988) is used to assess remote memory for visuospatial information. The subject views a map of the United States without state boundaries and is asked to locate cities on numbered dots. A patient with a right temporal lobectomy after removal of a glioblasma was impaired in the New map test but not the Fargo map test (Beatty et al., 1988), indicating anterograde but not retrograde amnesia. Likewise, patient EP with hippocampal damage caused by herpes simplex encephalitis was impaired in topographical memory of his current area of residence but not of places learned long ago, a result which underlines the role of this brain region in acquiring new information (Teng and Squire, 1999).

7. Spatial Short-Term and Working Memory

The Corsi block-tapping test and the spatial span of the Wechsler memory scale-revised version (WAIS-R, Wechsler, 1987) provide estimates of spatial short-term or working memory. The experimenter taps in quick succession a series of blocks (9 in the Corsi version, 7 in the Wechsler version) with the same shape but distributed in different locations on a table. The block-tapping test can also be administered in a computerized version (Morris et al., 1988). After a brief retention interval, the subject must reproduce the tap-blocking in the same order, amounting to a spatial version of the digit span subtest of the WAIS-R with non-verbal instead of verbal material. In either test, forward span may be dissociated from backward span. Unlike forward span, backward span implies the ability to manipulate information mentally and therefore depends on working as opposed to short-term memory.

Stroke patients with a visual field defect caused by lesions of either hemisphere had shorter spans on the Corsi block-tapping test than those without a visual field defect and patients with right hemisphere lesions were worse than those with left hemisphere lesions (De Renzi et al., 1977b). Corsi block-tapping performance was impaired in patients with lesions of either left or right anterior cortex or right posterior cortex (Vilkki and Holst, 1989). But when subjects were asked to judge how many blocks they could tap correctly, only patients with anterior lesions were impaired. These results underline the role of the prefrontal cortex in programming a goal-based search and the posterior parietal cortex in spatial short-term memory. Block-tapping can be adapted to evaluate long-term memory simply by augmenting the number of blocks to be remembered. Indeed, supra-span memory of Corsi block-tapping was deficient in a patient with diencephalic amnesia (Nichelli et al., 1988). More recent methods of evaluating spatial short-term memory include computer-generated images such as the Four mountains test, comprising landscapes shown from different viewpoints (Bird et al.,

2010), together with tests of object location (Alescio-Lautier et al., 2007; Nordahl et al., 2005).

Two other tests of spatial working memory include delayed response and delayed alternation (Oscar-Berman, 1991) adapted from research on animals (chapter 2). For both tests, the subject views a stimulus on the left or right side of a panel and chooses during the retention test which side the stimulus appeared in, the correct answer being on the same side for delayed response and on the opposite side for delayed alternation. The performance of both tasks was impaired in patients with lesions of prefrontal cortex or Korsakoff syndrome characterized by combined damage to prefrontal cortex and diencephalic nuclei (Oscar-Berman, 1991). As with the Corsi block-tapping test, lengthening the retention interval is used to assess spatial long-term memory. Spatial working memory has been theorized as a visuospatial sketch-pad distinct from articulatory processes, together forming the central executive (Della Scala and Logie, 1993).

Other tests of spatial working memory include the spatial recognition test, whereby the subject views on a computer screen a square at a certain location and then must select between two squares the one previously found at that location (Figure 7) (Owen et al., 1995). This task was impaired in a heterogenous group of patients with unilateral or bilateral lesions of the frontal lobe (Owen et al., 1995). In a similar task containing more choices, the subject views five identical figures and a different one at a specific location and must select the figure found at that location (Figure 8) (Owen et al., 1995). This task was impaired in a heterogenous group with unilateral or bilateral lesions of the frontal lobe along with groups with unilateral temporal lobe ablations with or without hippocampal-amygdala damage as a result of intractable epilepsy (Owen et al., 1995). Another spatial working memory test comprises searching through boxes on a computer screen to collect blue tokens inside. When a blue token is found, the box must not be opened a second time. This task was impaired in a heterogenous group of patients with unilateral or bilateral lesions of the frontal lobe, as well as those with unilateral medial temporal lobe ablations as a result of intractable epilepsy (Owen et al., 1990, 1995).

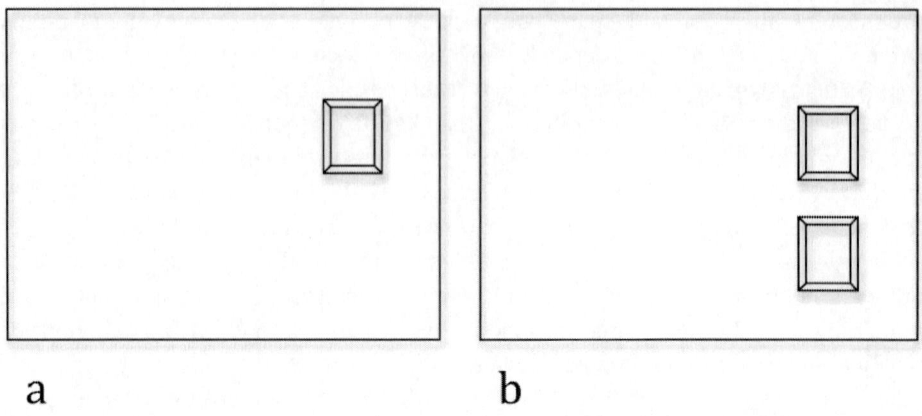

Figure 7. The two-choice spatial recognition test: the subject first views a square (a) and then selects between two squares (b), the correct answer being the one found at the previous location (Owen et al., 1995).

In the executive golf task (Morris et al., 1988), the subject must putt a ball inside a hole on a computer screen without returning to the same one twice. This task was impaired in patients with right temporal lobe ablations as a result of intractable epilepsy (Feigenbaum et al., 1996). The task was also impaired in patients with left or right frontal lobe ablations as a result of intractable epilepsy or other causes (Miotto et al., 1996). The structured golf task is a variation of the previous test in which the subject views pairs of holes and must select the novel one (Feigenbaum et al., 1996). This task was also impaired in patients with right temporal lobe ablations as a result of intractable epilepsy (Feigenbaum et al., 1996). The rotating spatial memory task uses allocentric information (Feigenbaum et al., 1996). The subject views holes to find a treasure kept within. After the treasure is found, the subject must avoid selecting the same hole a second time while the panel rotates the holes. As with the golf tasks, the rotating spatial memory task was impaired in patients with right temporal lobe ablations (Feigenbaum et al., 1996).

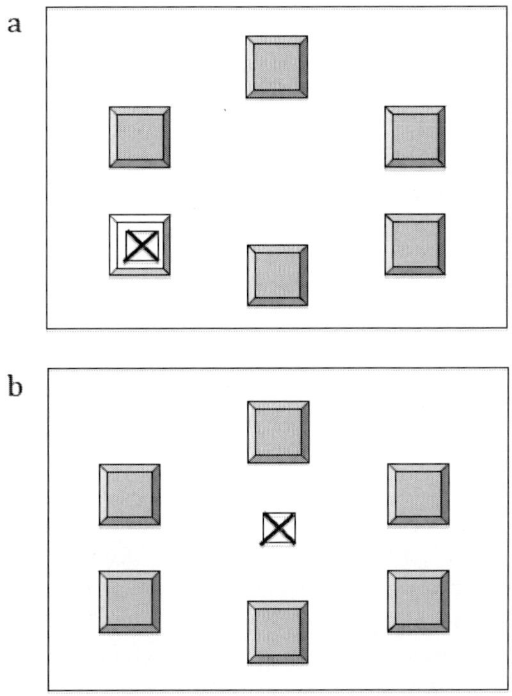

Figure 8. The six-choice spatial recognition test: the subject first views five identical figures and a different one at a specific location (a). He then must select between six choices (b) the previous location of the different square (Owen et al., 1995).

8. Spatial Long-Term Memory

Spatial long-term memory is usually evaluated by asking the subject to draw figures on sheets of paper after a retention interval of at least thirty minutes (Bear et al., 1996; Lezak, 1983), such as the Rey complex figure and the Benton visual retention test (Benton et al., 1983), as well as visual reproduction in the non-verbal portion of the Wechsler memory scale

(Wechsler, 1945). Other spatial memory tests containing geometric figures have been designed (Figure 9) (Jones-Gotman, 1986). Another option is to use three-dimensional objects spread on a tray (Smith, 1988). Spatial long-term memory tests are generally more sensitive to lesions located in the right as opposed to the left hemisphere (Lezak, 1983). Deficient recall of the Rey complex figure has been documented in patients with transient global amnesia (Kritchevsky et al., 1988). Similar to the verbal version of the Wechsler memory scale, paired-associate learning tests are available with figures (Goldstein et al., 1988). The difference score between spatial and verbal paired-associate learning distinguished patients with right as opposed to left temporal lobectomy (Goldstein et al., 1988).

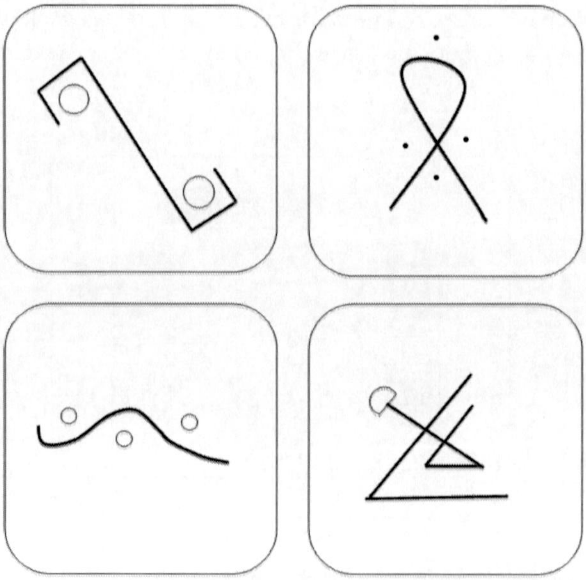

Figure 9. Spatial memory tests with four geometric figures used as stimuli to be remembered (Jones-Gotman, 1986).

Many case reports indicate the susceptibility of anterograde amnesia for visuospatial material following bilateral or right-sided damage to the medial temporal lobe. The first case of this kind in the history of neurology, HM, characterized by bilateral ablation of the medial temporal lobe to alleviate untractable seizures, was impaired in memorizing the location of three-dimensional objects spread on a tray (Smith, 1988). Anterograde amnesia was also documented in RB, characterized by bilateral ischemic lesions circumscribed to the CA4 subregion of the hippocampus (Squire, 1986). This patient had an IQ of 111 and a Weschsler memory score of 91, indicating a selective disturbance in memory. He was impaired in recall of the Rey complex figure (Squire, 1986). Moreover, anterograde amnesia was documented in EP (IQ=103) suffering from bilateral lesions caused by viral encephalitis and circumscribed to perirhinal, entorhinal, and parahippocampal cortices as well as hippocampus and amygdala (Stefanacci et al., 2000). He was impaired in recall of the Rey complex figure and the visual reproduction of figures of the Wechsler memory scale-revised version (Stefanacci et al., 2000). Even a small lesion of the right hippocampus caused a severe retention deficit of the Rey complex figure when examined in a single patient (Woods et al., 1982). This test is also sensitive to right anterior temporal lobectomy following intractable epilepsy (Jones-Gotman,

1986). A dissociation occurred between deficient retention of the Rey complex figure and intact Corsi block-tapping in a patient with bilateral lesions of the fornix after tumor removal, a result which indicates a selective deficit of long-term as opposed to short-term memory (Calabrese et al., 1995). These data along with experiments with verbal items emphasize the importance of the medial temporal lobe in the consolidation of new information into long-term memory, as this structure lies downstream of neocortical association areas involved in language, praxis, and shape recognition (Squire and Zola-Morgan, 1991).

In addition to lesions of the medial temporal cortex, anterograde amnesia is a frequent symptom arising from damage to the diencephalon, especially in the right hemisphere. Indeed, patients with infarcts in the right thalamus had worse retention scores of Benton figures than those on the left (Wallesch et al., 1983). Nevertheless, case reports indicate deficits in retention of the Rey complex figure and Benton figures following non-hemorrhagic infarcts both to right and left thalamus (Graff-Radford et al., 1984). The two tests were also affected by bilateral thalamic infarcts (Graff-Radford et al., 1990). Moreover, retention of Benton figures was impaired in two patients with bilateral thalamic infarcts whose performance on the Corsi block-tapping test was in the normal range, indicating a specific deficit in long-term memory (von Cramon et al., 1985). Retention of the Rey complex figure was deficient after bilateral infarcts of the inferior thalamic peduncle and the mamillary bodies (Malamut et al., 1992) or after bilateral removal of the mamillary bodies following a cystic craniophayngioma, visual reproduction of the Wechsler memory scale being also impaired (Tanaka et al., 1997). In addition, memory of the Rey complex figure was impaired after a selective infarct to the right anterior thalamus (Rousseaux et al., 1991).

As seen with vascular cases, patients with Korsakoff syndrome caused by chronic alcohol abuse and the resulting thiamine depletion, which causes lesions of the limbic thalamus and mamillary bodies, were impaired in recall of the Rey complex figure (Janowsky et al., 1989; Squire and Shimamuta, 1986). In contrast, retention of the Rey complex figure and Benton figures was intact in two patients with non-hemorrhagic infarcts destroying less than 15% of the dorsomedial thalamus (Kritchevsky et al., 1987).

In addition to damage to medial temporal cortex and diencephalon, anterograde amnesia is a frequent symptom arising from damage to the basal forebrain. Indeed, the recall of Benton figures was impaired in five patients with basal forebrain damage caused by rupture of anterior cerebral or anterior communicating arteries or by the resection of an arteriovenous malformation (Damasio et al., 1985b). In three of the four evaluated patients, recall of the Rey complex figure was also impaired. Likewise, both tests were deficient in a global amnesic patient with bilateral basal forebrain and temporal lobe damage caused by herpes simplex encephalitis (Damasio et al., 1985a). A more discrete bilateral lesion in the diagonal band of Broca caused by resection of low-grade glioma, with septal nuclei and basal nucleus of Meynert remaining intact, impaired retention of the Rey complex figure and visual reproduction of the Wechsler memory scale (Morris et al., 1992).

These results support the existence of a cortico-subcortical network underlying long-term memory, which includes medial temporal lobe and its afferents from medial septum, diagonal band of Broca, and basal nucleus of Meynert as well as efferents towards anterior thalamus and mamillary bodies projecting to the ventromedial prefrontal cortex.

REFERENCES

Alescio-Lautier, B; Michel, BF; Herrera, C; Elahmadi, A; Chambon, C; Touzet, C; Paban, V. Visual and visuospatial short-term memory in mild cognitive impairment and Alzheimer disease: role of attention. *Neuropsychologia,* 2007, 45, 1948-60.

Barra, J; Marquer, A; Joassin, R; Reymond, C; Metge, L; Chauvineau, V; Pérennou, D. Humans use internal models to construct and update a sense of verticality. *Brain,* 2010, 133, 3552-63.

Bear, MF; Connors, BW; Paradiso, MA. *Neuroscience: exploring the brain.* Baltimore: Williams & Wilkins, pp 514-545, 1996.

Beatty, WW. The Fargo Map Test: a standardized method for assessing remote memory for visuospatial information. *J. Clin. Psychol.,* 1988, 44, 61-7.

Beatty, WW; Tröster, AI. Gender differences in geographical knowledge. *Sex Roles,* 1987, 16, 565-90.

Beatty, WW; MacInnes, WD; Porphyris, HS; Tröster, AI; Cermak, LS. Preserved topographical memory following right temporal lobectomy. *Brain Cogn.,* 1988, 8, 67-76.

Benton, AL; Hamsher K de S; Varney NR, Spreen, O. *Contributions to neuropsycholical assessment: a clinical manual.* Oxford: Oxford University Press, 1983.

Berker, E; Smith, A. Diaschisis, site, time and other factors in raven performances of adults with focal cerebral lesions. *Int. J. Neurosci.,* 1988, 38, 267-85.

Bird, CM; Chan, D; Hartley, T; Pijnenburg, YA; Rossor, MN; Burgess, N. Topographical short-term memory differentiates Alzheimer's disease from frontotemporal lobar degeneration. *Hippocampus,* 2010, 20, 1154-69.

Botez, MI; Botez, T; Olivier, M. Parietal lobe syndromes. In: PJ Vinken, BW Bruyn, GL Klawans (eds) *Handbook of clinical neurology.* Elsevier: Amsterdam, pp 63-85, 1985.

Bowen, FP; Hoehn, MM; Yahr, MD. Parkinsonism, alterations in spatial orientation as determined by a route-walking test. *Neuropsychologia,* 1972, 10, 355-61.

Butters, N; Soeldner, C; Fedio, P. Comparison of parietal and frontal lobe spatial deficits in man: extrapersonal vs personal (egocentric) space. *Percept. Motor Skills,* 1972, 34, 27-34.

Calabrese, P; Markowitsch, HJ; Harders, AG; Scholz, M; Gehlen, W. Fornix damage and memory: a case report. *Cortex,* 1995, 31, 555-64.

Chouinard, S; Brière, ME; Rainville, C; Godbout, R. Correlation between evening and morning waking EEG and spatial orientation. *Brain Cogn.,* 2003, 53, 162-5.

Cohen, RJ; Montague, P; Nathanson, LS; Swerdlick, ME. *Psychological testing.* Mountain View: Mayfield Publishing, pp 217-63, 1988.

Costa, LD. Interset variability on the raven coloured progressive matrices as an indicator of specific ability deficit in brain-lesioned patients. *Cortex,* 1976, 12, 31-40.

Cytowic, RE. *The neurological side of neuropsychology.* Cambridge: MIT, pp 417-58, 1996.

Damasio, AR; Eslinger, PJ; Damasio, H; Van Hoesen, GW; Cornell, S. Multimodal amnesic syndrome following bilateral temporal and basal forebrain damage. *Arch. Neurol.,* 1985a, 42, 252-9.

Damasio, AR; Graff-Radford, NR; Eslinger, PJ; Damasio, H; Kassell, N. Amnesia following basal forebrain lesions. *Arch. Neurol.,* 1985b, 42, 263-71.

Della Scala, S; Logie, RH. When working memory does not work: the role of working memory in neuropsychology. In: H Spinnler, F Boller (eds) *Handbook of Neuropsychology,* vol 8. Elsevier, Amsterdam, pp 1-62. 1993.

de Renzi, E; Faglioni, P; Villa, P. Topographical amnesia. *J. Neurol. Neurosurg. Psychiatry,* 1977a, 40, 498-505.

de Renzi, E; Faglioni, P; Previdi, P. Spatial memory and hemispheric locus of lesion. *Cortex,* 1977b, 13, 424-33.

Fehrenbach, RA; Wallesch, C-W; Claus, D. Neuropsychologic findings in Friedreich's ataxia. *Arch. Neurol.,* 1984, 41, 306-8.

Feigenbaum, JD; Polkey, CE; Morris, RG. Deficits in spatial working memory after unilateral temporal lobectomy in man. *Neuropsychologia,* 1996, 34, 163-76.

Goldstein, LH; Canavan, AGM; Polkey, CE. Verbal and abstract designs paired associate learning after unilateral temporal lobectomy. *Cortex,* 1988, 24, 41-52.

Graff-Radford, NR; Eslinger, PJ; Damasio, AR; Yamada, T. Nonhemorrhagic infarction of the thalamus: behavioral, anatomic, and physiologic correlates. *Neurology,* 1984, 34, 14-23.

Graff-Radford, NR; Tranel, D; Van Hoesen, GW; Brandt, JP. Diencephalic amnesia. *Brain,* 1990, 113, 1-25.

Hort, J; Laczo, J; Vyhnalek, M; Bojar, M; Bures, J; Vlcek, K. Spatial navigation deficit in amnestic mild cognitive impairment. *Proc. Natl. Acad. Sci. USA,* 2007, 104, 4042-7.

Janowsky, JS; Shimamura, AP; Kritchevsky, M; Squire, LR. Cognitive impairment following frontal lobe damage and its relevance to human amnesia. *Behav. Neurosci.,* 1989, 103, 548-60.

Jones-Gotman, M. Right hippocampal excision impairs learning and recall of a list of abstract designs. *Neuropsychologia,* 1986, 24, 659-70.

Kertesz, A; Dobrowolski, S. Right-hemisphere deficit, lesion size and location. *Int. J. Neuropsychol.,* 1981, 3, 283-99.

Kritchevsky, M; Graff-Radford, NR; Damasio, AR. Normal memory after damage to medial thalamus. *Arch. Neurol.,* 1987, 44, 959-62.

Kritchevsky, M; Squire, LR; Zouzounis, JA. Transient global amnesia: characterization of anterograde and retrograde amnesia. *Neurology,* 1988, 38, 213-9.

Lhermitte, F; Derouresne, J; Signoret, J-L. Analyse neuropsychologique du syndrome frontal. *Rev. Neurol.,* 1972, 127, 415-40.

Lezak, MD. *Neuropsychological assessment.* 2nd edition. Oxford: Oxford University Press, 1983.

Majerová, V, Kalinčík, T; Laczó, J; Vyhnálek, M; Hort, J; Bojar, M; Růžička, E; Roth, J. Disturbance of real space navigation in moderately advanced but not in early Huntington's disease. *J. Neurol. Sci.,* 2012, 312, 86-91.

Malamut, BL; Graff-Radford, N; Chawluk, J; Grossman, RI; Gur, RC. Memory in a case of bilateral thalamic infarction. *Neurology,* 1992, 42, 163-9.

McCarthy, RA; Evans, JJ; Hodges JR. Topographic amnesia: spatial memory disorder, perceptual dysfunction, or category specific semantic memory impairment? *J. Neurol. Neurosurg. Psychiatry,* 1996, 60, 318-25.

Milner, B. Visually-guided maze learning in man: effects of bilateral hippocampal, bilateral frontal and unilateral cerebral lesions. *Neuropsychologia,* 1965, 3, 317-38.

Milner, B; Corkin, S; Teuber H-L. Further analysis of the hippocampal amnesic syndrome: 14-year follow-up study of HM. *Neuropsychologia,* 1968, 6, 215-34.

Miotto, EC; Bullock, P; Polkey, CE; Morris, RG. Spatial working memory and strategy formation in patients with frontal lobe excisions. *Cortex,* 1996, 32, 613-30.

Monacelli, AM; Cushman, LA; Kavcic, V, Duffy, CJ. Spatial disorientation in Alzheimer's disease: the remembrance of things passed. *Neurology,* 2003, 61, 1491-7.

Money, J. *A standardized road map test of direction sense*. San Rafael CA: Academic Theapy Publications, 1976.

Morris, MK; Bowers, D; Chatterjee, A; Heilman, KM. Amnesia following a discrete basal forebrain lesion. *Brain,* 1992, 115, 1827-47.

Morris, RG; Downes, JJ; Sahakian, BJ; Evenden, JL; Heald, A; Robbins, TW. Planning and spatial working memory in Parkinson's disease. *J. Neurol. Neurosurg. Psychiatry,* 1988, 51, 757-66.

Nichelli, P; Bahmannian-Behhahani, G; Gentilini, M; Vecchi, A. Preserved memory abilities in thalamic amnesia. *Brain,* 1988, 111, 1337-53.

Nordahl, CW; Ranganath, C; Yonelinas, AP; DeCarli, C; Reed, BR; Jagust, WJ. Different mechanisms of episodic memory failure in mild cognitive impairment. *Neuropsychologia,* 2005, 43, 1688-97.

Oscar-Berman, M. Clinical and experimental approaches to varieties of memory. *Int. J. Neurosci.,* 1991, 58, 135-50.

Osterrieth, PA. Le test de copie d'une figure complexe. *Arch. Psychol.,* 1944, 30, 206-356.

Ott, BR; Festa, EK; Amick, MM; Grace, J; Davis, JD; Heindel, WC. Computerized maze navigation and on-road performance by drivers with dementia. *J. Geriatr. Psychiatry Neurol.,* 2008, 21, 18-25.

Owen, AM; Downes, JJ; Sahakian, BJ; Polkey CE; Robbins TW. Planning and spatial working memory following frontal lobe lesions in man. *Neuropsychologia,* 1990, 28, 1021-34.

Owen, AM; Sahakian, BJ; Semple, J; Polkey, CE; Robbins, TW. Visuo-spatial short-term recognition memory and learning after temporal lobe excisions, frontal lobe excisions or amygdalo-hippocampectomy in man. *Neuropsychologia,* 1995, 33, 1-24.

Pengas, G; Patterson, K; Arnold, RJ; Bird, CM; Burgess, N; Nestor, PJ. Lost and found: bespoke memory testing for Alzheimer's disease and semantic dementia. *J. Alzheimers Dis.,* 2010, 21, 1347-65.

Porteus, SD. *Porteus maze tests: fifty years of application*. Palo Alto: Pacific Books, 1965.

Raven, JC; Court, JH; Raven, J. *Manual for Raven's progressive matrices and vocabulary scales-standard progressive matrices*. London: HK Lewis, 1984a.

Raven, JC; Court, JH; Raven, J. *Manual for Raven's progressive matrices and vocabulary scales-coloured progressive matrices*. London: HK Lewis, 1984b.

Rey, A. *L'examen clinique en psychologie*. Paris: Presses Universitaires de France, 1965.

Rousseaux, M; Kassiotis, P; Signoret, JL; Cabaret, M; Petit H. (Amnesic syndrome caused by limited infarction in the right anterior thalamus) (in French). *Rev. Neurol.,* 1991, 147, 809-18.

Schneider, GE. Two visual systems. *Science,* 1969, 163, 895-902.

Smith, ML. Recall of spatial location by the amnesic patient H.M. *Brain Cogn.,* 1988, 7, 178-83.

Squire, LR. Mechanisms of memory. *Science,* 1986, 232, 1612-9.

Squire, LR; Shimamura, AP. Characterizing amnesic patients for neurobehavioral study. *Behav. Neurosci.,* 1986, 100, 866-77.

Squire, LR; Zola-Morgan, S. The medial temporal lobe memory system. *Science,* 1991, 253, 1380-6.

Stefanacci, L; Buffalo, EA; Schmolck, H; Squire LR. Profound amnesia after damage to the medial temporal lobe: A neuroanatomical and neuropsychological profile of patient E.P. *J. Neurosci.,* 2000, 20, 7024-36.

Teng, E; Squire, LR. Memory for places learned long ago is intact after hippocampal damage. *Nature,* 1999. 400. 675-7.

Teuber, HL; Mishkin, M. Judgment of visual and postural vertical after brain injury. *J. Psychol.,* 1954, 38, 161-75.

Tranel, D; Vianna, E; Manzel, K; Damasio, H; Grabowski, T. Neuroanatomical correlates of the Benton Facial Recognition Test and Judgment of Line Orientation Test. *J. Clin. Exp. Neuropsychol.,* 2009, 31, 219-33.

Uc, EY; Rizzo, M; Anderson, SW; Shi, Q; Dawson, JD. Driver route-following and safety errors in early Alzheimer disease. *Neurology,* 2004, 63, 832-7.

Ungerleider, LG; Mishkin, M. Two cortical visual systems. In: DJ Ingle, MA Goodale, RJW Mansfield (eds) *Analysis of visual behavior.* Boston: MIT Press, 1982.

Vilkki, J; Holst, P. Deficient programming in spatial learning after frontal lobe damage. *Neuropsychologia,* 1989, 27, 971-6.

von Cramon, DY; Hebel, N; Schuri, U. A contribution to the anatomical basis of thalamic amnesia. *Brain,* 1985, 108, 993-1008.

Wallesch, CW; Kornhuber, HH; Kunz, T; Brunner, RJ. Neuropsychological deficits associated with small unilateral thalamic lesions. *Brain,* 1983, 106, 141-52.

Wallesch, CW; Karnath, HO; Papagno, C; Zimmermann, P; Deuschl, G; Lücking, CH. Parkinson's disease patient's behaviour in a covered maze learning task. *Neuropsychologia,* 1990, 28, 839-49.

Wechsler, D. A standardized memory scale for clinical use. *J. Psychol.,* 1945, 19, 87-95.

Wechsler, D. *Wechsler memory scale-revised edition.* New York: The Psychological Corporation, 1987.

Woods, BT; Schoene, W; Kneisley, L. Are hippocampal lesions sufficient to cause lasting amnesia? *J. Neurol. Neurosurg. Psychiatry,* 1982, 45, 243-7.

In: Spatial Orientation in Humans and Animals
Editor: Robert Lalonde
ISBN: 978-1-63463-001-6
© 2014 Nova Science Publishers, Inc.

Chapter 2

SPATIAL TASKS USED IN ANIMAL MODELS OF BRAIN DISORDERS

Robert Lalonde[1,*] and Catherine Strazielle[2,#]
[1]Université de Rouen, Dépt. Psychologie, Mont-Saint-Aignan Cedex France
[2]Université de Lorraine, Laboratoire "Stress, Immunité, Pathogènes" and Service de Microscopie Electronique, Faculté de Médecine and CHU de Nancy,Vandoeuvre-les-Nancy, France

1. MORRIS WATER MAZE

The analysis of brain lesions on maze learning has a history dating at least to Karl Lashley's experiments in the 1920s (Bear et al., 1996; Lashley, 1929). It was quickly noticed that laboratory rats and mice are efficient maze learners because of their ability to recognize points of reference relative to a goal. Likewise, in the wild, nightbirds such as the indigo bunting are able to orient themselves by looking at the position of stars (Emlen, 1975) and bees at the position of the sun (von Frisch, 1967). The effects of brain lesions may be examined either during acquisition of a novel task or during performance of an already learned task. Cortical lesions are often made by aspiration, subcortical ones by radio-frequency or direct current. Relative to the electrolytic method, excitotoxic lesioning with ibotenate and other substances delineates more precisely the impact of grey matter loss while sparing fibers of passage. Temporary inactivation of a brain area may be estimated by injections of lidocaine.

One of the most often used spatial learning task is the Morris water maze (Figure 1), where mice or rats placed inside a pool of water from start locations varying at all four cardinal points learn to escape by reaching either a submerged or a visible platform (Morris et al., 1982). To keep the platform hidden, the water is made opaque by milk, non-toxic paint, or plastic beads (Cain et al., 1993). While the hidden platform subtask reflects learning on the basis of extra-maze cues, the visible platform one reflects visuomotor guidance towards a directly observable target. The hidden platform task concerns reference memory in that the goal is always found in the same location, unlike working memory tasks when the goal

* Email: robert.lalonde@univ-rouen.fr
# Email: catherine.strazielle@univ-lorraine.fr

changes from one trial to the next (Walker and Olton, 1979). Poor visible platform performance may be due to non-cognitive factors such as a loss in motivation or visual acuity. Unlike land-based tasks where a win-shift strategy prevails, water maze tasks cause rats to adopt a win-stay strategy, presumably because the goal is felt to be more urgent (fear of drowning) (Means, 1988). A deficiency in place as opposed to cued learning is usually interpreted as a specific effect on cognition. However, minor defects in visual acuity caused by visual deprivation during development induced the same pattern (Prusky et al., 2000), and so care should be taken to examine more extensively the impact of brain lesions on visual acuity.

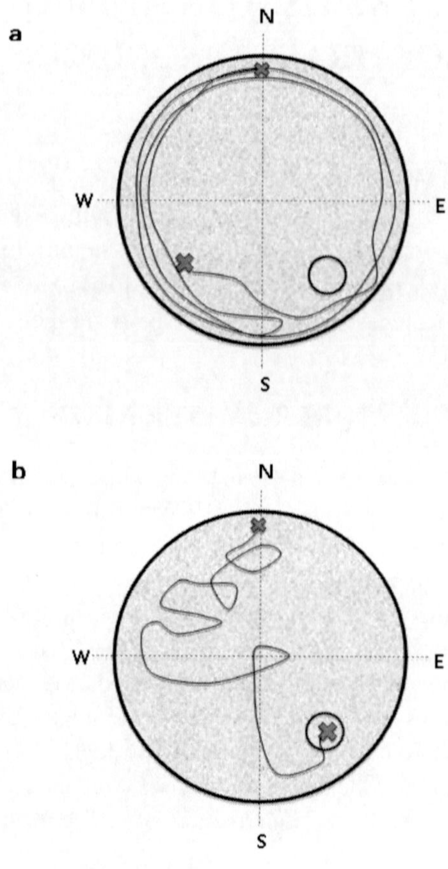

Figure 1. Diagram of the Morris water maze (Morris et al., 1982). In the first example (a), the animal (x) is released on the northern (N) part of the pool and mostly swims around the perimeter without finding the escape platform located in the southeast (SE) quadrant. In the next example (b), the animal succeeds in reaching the platform via a roundabout route.

After unsuccessful trials, the animals are usually carried to the platform (corrected trials). When uncorrected trials are used, the animal may learn to give up searching and look for the experimenter's hand instead of the platform. To ensure that the hidden platform task comprises original training on naive animals, the visible platform task is often given afterwards. It is sometimes given before when it is doubtful whether the animals will achieve an acceptable learning curve, notably due to the suboptimal genetic background of the mice.

In both cases, the visible platform is usually placed in a different quadrant than the hidden platform. Otherwise, the learning underlying each task is confounded. To maximize performances, hypothermia should be avoided by extending intertrial intervals to at least 13 min for water temperatures ranging between 20° and 24°C (Iivonen et al., 2003). Usually after acquisition, sometimes throughout, one or more probe trials are given with the hidden platform removed from the pool. Thereby, the accuracy of spatial long-term memory is shown whenever the animal swims in the vicinity of where the platform used to be.

D'Hooge and De Deyn (2001) summarized the sensitivity of the Morris water maze task to surgical lesions of various brain areas as well as the consequences of vascular anomalies, neurotrauma, and developmental or metabolic disorders. McNamara and Skelton (1993) more specifically summarized the sensitivity of this test to drugs.

1.1. Hidden Platform Subtask

The hidden platform subtask depends on visual input derived from posterior neocortex (Table 1), as shown by deficits after excitotoxic lesions of primary visual cortex plus posterior parietal areas in rats (Hoh et al., 2003) or anteromedial extrastriate visual cortex (Espinoza et al., 1999). The role of the occipito-parietal projection is crucial in this task, as demonstrated by delayed acquisition after aspiration lesions of the posterior parietal cortex (DiMattia and Kesner, 1988a; Kesner et al., 1991; Kolb and Walkey, 1987; Kolb et al., 1983, 1994; Save and Moghaddam, 1996). Spatial information is eventually conveyed to the medial temporal lobe (chapter 3). Indeed, the acquisition of the task was likewise delayed after aspirations lesions of the entorhinal cortex in rats (Nagahara et al., 1995) or electrolytic lesions of the entorhinal cortex in mice (Hardman et al., 1997). However, no detrimental effect was found after excitotoxic lesions of perirhinal, postrhinal, or combined perirhinal, postrhinal, and entorhinal cortex in rats (Burwell et al., 2004), perhaps because such massive damage led to compensatory actions from other brain regions or because of the use of an unsually brief intertrial interval (30 sec).

The entorhinal cortex is the main source of input to the hippocampus (chapter 3). The importance of the hippocampus on place learning in rats is amply demonstrated by findings indicating slowed acquisition relative to sham-operated controls after aspiration (Morris et al., 1982), electrolytic (DiMattia and Kesner, 1988b; Sutherland et al., 1983), excitotoxic (Marston et al., 1993; Richmond et al., 1999; Sutherland et al., 1983), or colchicine-induced (Sutherland et al., 1983) lesions. Damage to hippocampal subregions is often sufficient to cause a deficit, as indicated by excitotoxic lesions of CA1 (Stubley-Weatherly et al., 1996), CA3 (Brun et al., 2002; Stubley-Weatherly et al., 1996), or subiculum (Morris et al., 1990). When dorsal versus ventral hippocampal damage was compared, excitoxic lesions of only the latter caused a deficit (Bannerman et al. 1999; Moser et al. 1995). However, no deficit in either subregion was found after excitotoxic lesions in a third study (Richmond et al., 1999). Nevertheless, the overall results provide behavioral confirmation of electrophysiological data indicating the existence of place cells (O'Keefe and Dostrovsky, 1971) encoding allocentric information (Feigenbaum and Rolls, 1991) and dependent on movement preparation (Foster et al., 1989) in the hippocampus. Thus, this type of spatial learning depends on medial temporal lobe structures receiving visual input from posterior parietal and temporal association cortex (Pandya and Yeterian, 1984; Poucet, 1993).

Table 1. The effects of brain lesions on acquisition of the hidden platform task of the Morris water maze in rats as well as probe and visible platform tasks

Brain regions	Hidden platform	Probe trial	Visible platform	References
Posterior neocortex	↓	?	?	Espinoza et al., 1999; Hoh et al., 2003
Posterior parietal cortex	↓	?	No effect	DiMattia and Kesner, 1988a; Kesner et al., 1991; Kolb and Walkey, 1987; Kolb et al., 1983, 1994; Save and Moghaddam, 1996
Anterior or prefrontal neocortex	↓	?	No effect	Hoh et al., 2003; Kolb et al., 1982, 1983, 1994; Sutherland et al., 1982
Orbitofrontal cortex	↓	?	?	Kolb et al., 1983
Anterior cingulate cortex	↓	?	?	Sutherland et al., 1988; Warburton et al., 1998; Whishaw et al., 2001
Posterior cingulate cortex	↓	?	?	Sutherland et al., 1988; Vann et al., 2003
Entorhinal cortex	↓	↓	No effect	Nagahara et al., 1995
Hippocampus	↓	↓	↓	Burwell et al., 2004; DiMattia and Kesner, 1988b; Marston et al., 1993; Morris et al., 1982; Richmond et al., 1999; Sutherland et al., 1983
Medial septum	↓	↓	↓	Hagan et al., 1988; Kelsey and Landry, 1988; Miyamoto et al., 1987; Riekkinen et al., 1990a; Sutherland and Rodriguez, 1989
Amygdala	No effect	?	?	Sutherland and McDonald, 1990
Nucleus basalis magnocellularis	↓	No effect	No effect	Hagan et al., 1988; Mayo et al., 1988; Miyamoto et al., 1987; Riekkinen et al., 1990b
Anterior thalamus	↓	?	No effect	Sutherland and Rodriguez, 1989
Mamillary bodies	↓	?	No effect	Sutherland and Rodriguez, 1989
Superior colliculus	?	?	↓	Lines and Milner, 1985
Habenula	↓	?	?	Thornton and Davies, 1991

↓ slowed acquisition vs controls, ? not evaluated.

Afferent and efferent fibers to and from the hippocampus are also involved, as indicated by delayed acquisition of the hidden platform task after knife cuts (Skelton and McNamara, 1992) or electrolytic lesions (Schenk and Morris, 1985) into perforant path and after knife cuts (Segal et al., 1989) or electrolytic lesions (de Bruin et al., 2001; Sutherland and Rodriguez, 1989) of the fimbria-fornix. On the contrary, no such deficit was found after electrolytic lesions of the amygdala (Sutherland and McDonald, 1990).

Other structures included in spatial circuitry are hippocampal afferents originating from the basal forebrain. Indeed, the acquisition of the hidden platform task was slowed down after electrolytic (Segal et al., 1989) or excitotoxic (Marston et al., 1993) lesions of the septum/diagonal band of Broca or lesions more restricted to the medial septum by electrolytic (Kelsey and Landry, 1988; Miyamoto et al., 1987; Sutherland and Rodriguez, 1989) or excitotoxic (Hagan et al., 1988; Riekkinen et al., 1990a) means. However, in other studies, no detrimental effect was discerned after electrolytic (Decker et al., 1992), excitotoxic (Decker et al., 1992), or colchicine-induced (Barone et al., 1991) lesions of the medial septum. This discrepancy may be explained by the extent of damage to horizontal or vertical limbs of the diagonal band of Broca or variable recovery periods after lesioning. Although some studies showed delayed acquisition of the task after excitotoxic (Mayo et al., 1988; Miyamoto et al., 1987; Riekkinen et al., 1990b) or colchicine-induced (Mundy et al., 1990) lesions of nucleus basalis magnocellularis, Hagan et al. (1988) showed no such effect after excitotoxic lesions. Moreover, no effect was reported after selective cholinergic lesions of either medial septum or nucleus basalis magnocellularis (Baxter et al., 1995) or both (Baxter et al., 1996) with 192 IgG-saporin, an immunoneurotoxin composed of an antibody to the low affinity nerve growth factor receptor. It should be noted that lesions aimed at specific neurochemical systems are often more liable to compensatory effects arising from other neurotransmitter systems than more generalized methods. Indeed, cholinergic lesions caused by intraventricular administration of 192 IgG-saporin slowed down acquisition of the task in mice (Berger-Sweeney et al., 2001; Moreau et al., 2008), an effect attributable to more complete damage to cholinergic neurons throughout the brain, including those of the brainstem (chapter 3). Other subcortical regions involved in acquisition include the habenula, as seen after electrolytic lesions of this structure (Thornton and Davies, 1991).

Hippocampal output towards the diencephalon also takes part in the spatial network underlying the task. Indeed, acquisition was slowed down after electrolytic lesions of anterior thalamus (Sutherland and Rodriguez, 1989) or mamillary bodies (Sutherland and Rodriguez, 1989) and excitotoxic lesions of dorsomedial thalamus (Dolleman-van der Weel et al., 2009), structures conveying spatial information towards the prefrontal cortex (Jankowsky et al., 2013; Kolb, 1984). The importance of the prefrontal cortex on this task is demonstrated by the delayed acquisition found after large lesions comprising Cg1+Cg3+infralimbic+Fr2 (Hoh et al., 2003; Kolb et al., 1994) or precentral+prelimbic+anterior cingulate (Kolb et al., 1982, 1983; Sutherland et al., 1982) regions, as well as smaller lesions comprising orbitofrontal (Kolb et al., 1983), anterior cingulate (Sutherland et al., 1988; Warburton et al., 1998; Whishaw et al., 2001), or posterior cingulate (retrosplenial) (Sutherland et al., 1988; Vann et al., 2003; Whishaw et al., 2001). However, no effect occurred after larger electrolytic lesions of medial prefrontal cortex comprising Fr2+dorsal anterior cingulate+prelimbic+infralimbic (de Bruin et al., 1994, 2001). These authors argued that the positive data cited above were caused by the use of uncorrected as opposed to corrected trials, leading rats to abandon effective search strategies. As mentioned above, the discrepancy may be also be caused by a paradoxically greater recovery following larger lesions, as seen in other spatial tasks such as delayed alternation (Irle, 1985).

The performance of the already acquired hidden platform subtask has more rarely been examined. It has been shown to be impaired after aspiration lesions of posterior parietal (DiMattia and Kesner, 1988a; Save and Moghaddam, 1996) or entorhinal (Nagahara et al., 1995) cortex, knife cuts into perforant path (Kirkby and Higgins, 1998) or the CA3 hippocampal subregion (Steffenach et al., 2002), and electrolytic lesions of fimbria-fornix, anterior thalamus, or mamillary bodies (Sutherland and Rodriguez, 1989), indicating the importance of these brain regions on retention.

1.2. Probe Trial

As in the hidden platform task, the probe trial was impaired after aspiration (Morris et al., 1982) or excitotoxic (Burwell et al., 2004; Marston et al., 1993) lesions of the hippocampus in rats, or else excitotoxic lesions limited to the CA3 subregion (Brun et al., 2002). The probe trial was also deficient after electrolytic (Miyamoto et al., 1987) or excitotoxic (Hagan et al., 1988; Marston et al., 1993) lesions of medial septum/diagonal band of Broca or aspirations lesions of the entorhinal cortex (Nagahara et al., 1995). Moroever, the probe trial was impaired in mice after electrolytic lesions of the entorhinal cortex (Hardman et al., 1997). On the contrary, the probe trial was intact after excitotoxic lesions of the retrosplenial cortex in rats (Vann et al., 2003) as well as excitotoxic lesions of nucleus basalis magnocellularis (Hagan et al., 1988; Miyamoto et al., 1987).

1.3. Visible Platform Subtask

The visible platform subtask was impaired after lesions of the superior colliculus (Lines and Milner, 1985), a result which confirms the importance of this brain region with respect to reaching a visible target, demonstrated in other paradigms (Schneider, 1969). In contrast, the performance of the visible platform subtask is generally spared after limbic lesions, but sometimes temporarily or more minimally affected relative to the hidden platform subtask. Indeed, the visible platform subtask was impaired in rats after aspiration lesions of the hippocampus only during the first few trials (Morris et al., 1982). Visible platform performance was also deficient in rats after electrolytic lesions of the fimbria-fornix (Eichenbaum et al., 1990) or excitotoxic lesions of the medial septum (Hagan et al., 1988). Likewise, the visible platform subtask was impaired in mice after electrolytic lesions of the entorhinal cortex (Hardman et al., 1997).

On the contrary, the visible platform subtask was not affected by aspiration lesions of posterior parietal (Kolb and Walkey, 1987), prefrontal (Cg1, Cg3, infralimbic, Fr2) (Kolb et al., 1994), or entorhinal (Nagahara et al., 1995) cortex, electrolytic lesions of fimbria-fornix, anterior thalamus, mamillary bodies (Sutherland and Rodriguez, 1989), or medial septum (Kelsey and Landry, 1988), cholinergic lesions of medial septum or nucleus basalis magnocellularis (Baxter et al., 1995), or excitotoxic lesions of nucleus basalis magnocellularis (Hagan et al., 1988).

1.4. Parameter Variations

The Morris water maze has been subject to several parameter variations, including a landmark version whereby the location of the escape platform is marked by a stimulus nearby (Kolb and Walkey, 1987; Save and Poucet, 2000). Aspiration (Kolb and Walkey, 1987; Kolb et al., 1994) or electrolytic (Save and Poucet, 2000) lesions of the posterior parietal cortex slowed down acquisition of this version, as well as aspiration lesions of prefrontal cortex (Cg1, Cg3, infralimbic, Fr2) (Kolb et al., 1994) or electrolytic lesions of the dorsal hippocampus (Save and Poucet, 2000). An egocentric version has been devised with variable start positions but with the escape platform still fixed at a constant location relative to the starting point (Save and Moghaddam, 1996). Aspiration lesions of the posterior parietal cortex impaired acquisition and performance of this version (Save and Moghaddam, 1996). Yet another variation comprises spatial discrimination learning with two platforms, one of which providing escape from the pool (Decker et al., 1992). Electrolytic but not excitotoxic lesions of the medial septum slowed down acquisition of this double platform version (Decker et al., 1992). The same impairment was found after electrolytic lesions of the fimbria-fornix but not the amygdala (Decker et al., 1995). In a path integration task, animals begin a trial from the same starting point while reaching the hidden platform at the same location in the light and then in the dark (Rochefort et al., 2011).

A matching-to-place version consists of two trials per day with escape platform locations changing from one day to the next from variable starting points on the same day, a task dependent on working as opposed to reference memory (Baxter et al., 1995; Brandeis et al., 1989; Whishaw, 1985). As with the standard version, rats appear to be better than mice (at least the C57BL strain) in this task (Whishaw, 1985). The acquisition of this version was slowed down after aspirations lesions of the hippocampus or subiculum (Morris et al., 1990), or else entorhinal cortex (Nagahara et al., 1995), knife cuts into fimbria-fornix (Turnbull et al., 1994), as well as selective cholinergic lesions of medial septum or nucleus basalis magnocellularis after administration of 192 IgG-saporin (Baxter et al., 1995). A water maze version with distracting visual stimuli has also been used (Thornton and Davies, 1991).

As with the hidden platform condition, some experimenters use a visible platform version with a constant-start location (Eichenbaum et al., 1990). Electrolytic lesions of the fimbria-fornix caused a defect in this version (Eichenbaum et al., 1990).

2. Multi-Unit Mazes

Mazes with multiple choice points along a path from the start box to the goal have the longest history in behavioral neuroscience, pioniered by Lashley (1929), originator of the Lashley III maze (Hoplight et al., 1996; O'Leary and Brown, 2013). The solution of these mazes relies on reference memory, since, like the Morris water maze, the goal always remains the same. A series of Hebb-Williams (1946) maze patterns were designed by Rabinovitch and Rosvold (1951) (Figure 2) and have often been used subsequently (Meunier et al., 1986; O'Leary and Brown, 2013; Pritchett and Mulder, 2004). Rats and mice are efficient in solving such maze problems motivated by food reward, water or shock escape, and access to a nesting site. Submammallian species such as ants can also solve them (Bonner, 1980; Schnerla,

1953). The number of errors, defined by entries into blind alleys, is the main measure used, along with latencies before reaching the goal. As with the Morris water maze, the animal may use allocentric (extra-maze) cues to find the goal. An egocentric version relying on intra-maze cues may be adopted by raising the maze walls at a sufficient height as to prevent the animal from seeing outside (Rogers and Kesner, 2006).

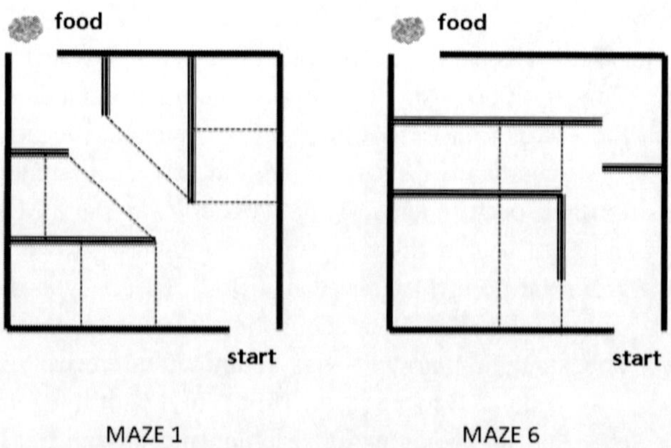

Figure 2. Diagram of two Hebb-Williams (1946) maze patterns of the Rabinovitch and Rosvold (1951) series: the animal is placed in the start box and must reach the goal box where food is found. On maze problem 1, the path is diagonal. On the more complex maze problem 2, the animal must turn right and then left. Crossing past the dotted lines is counted as an error.

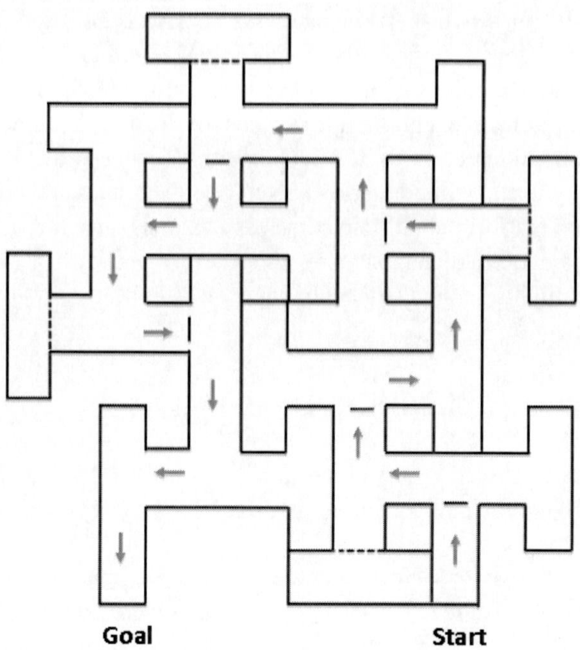

Figure 3. Diagram of a 14-unit maze (Kametani et al., 1993): the animal is placed in the start box and by turning left or right at each of 14 choice points must reach the goal box where food is found or where escape from footshocks is provided. The correct path is indicated by arrows.

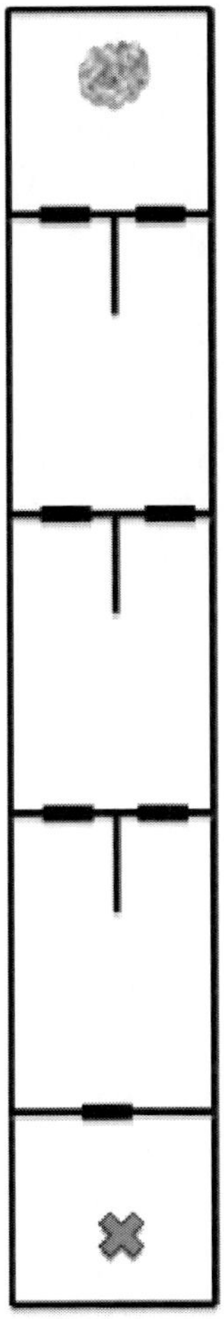

Figure 4. Diagram of a 3-unit maze (Buzsaki et al., 1982): the animal (x) is placed in the start box and must reach the goal where food is found by turning either to the left or the right, one of the doors being open (correct choice) and the other closed (error).

The acquisition of the 6-alley Lashley III maze task motivated by food reward was delayed after aspiration lesions of the hippocampus (Jackson and Strong, 1969) and medial frontal or posterior parietal cortex (Thomas and Weir, 1975). When all twelve problems of the Hebb-Williams series were presented to rats for food reward, acquisition was slowed down by aspiration lesions of the entire neocortex (Ain et al., 1969) or those confined to parietal cortex (Boyd and Thomas, 1977). When maze problems 1 and 6 (Kimble, 1963) or problem 11 (Kimble et al., 1982) were presented, the acquisition was slowed down after aspiration lesions of the hippocampus. Likewise, lidocaine injections in the CA1 subregion of the hippocampus or the medial prefrontal cortex impaired encoding and retention of the Hebb-Williams maze (Churchwell et al., 2010).

Other multiple-unit mazes include those with 14 (Figure 3), 9 (known as the Cincinnati maze) (Vorhees, 1987), 3 (Figure 4), or 2 units in Z (Lalonde et al., 1996) or double Y (Beninger et al., 1994; Mason et al., 1999) configurations. The acquisition of the 14-unit maze motivated by shock escape is sensitive to lesions of the medial septum (Kametani et al.,

1993). The acquisition of the 3-unit maze motivated by food reward was slowed down after electrolytic lesions of the medial fimbria (Buzsaki et al., 1982). In a shock escape paradigm, acquisition was slowed down after electrolytic lesions of parietal cortex, external globus pallidus, lateral substantia nigra, dorsal hippocampus, anterior thalamus, medial raphe, or pontine reticular formation (Thompson et al., 1986). Performance of the preoperatively acquired task was impaired after electrolytic lesions of posterior neocortex, hippocampus, anterior thalamus, and brainstem reticular formation (Thompson, 1974). In the Maier 3-table reasoning task (Herrmann et al., 1985a,b), rats are fed on one of 3 tables interconnected by a Y-shaped runway and then placed on one of the two others. Their task is to return to the table where they had just been fed. The starting point and the goal changes from one day to the next. The acquisition of this task was sensitive to electrolytic lesions of medial frontal cortex (Herrmann et al., 1985a) or septum (Herrmann et al., 1985b).

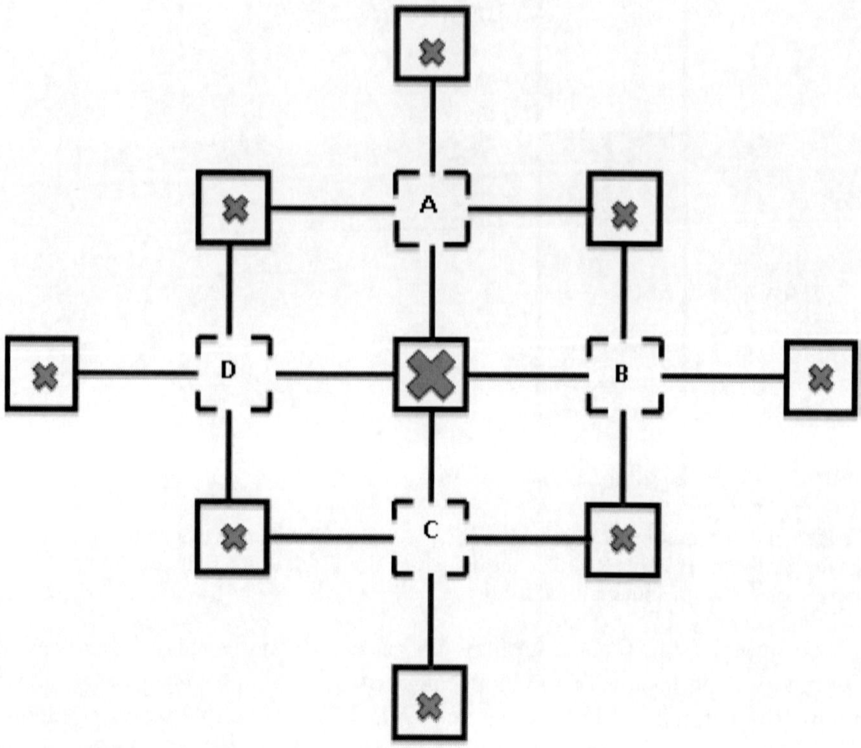

Figure 5. Diagram of an elevated maze with 4 start boxes (A, B, C, or D) and 8 goal boxes (X) for testing reference memory (Walker and Olton, 1984): the animal is placed inside one of the start boxes and must reach the goal box by avoiding all other boxes.

Maze configurations with multiple start boxes and goal boxes have been designed (Figure 5) (Walker and Olton, 1984). In a working memory version with 2 start boxes and 2 goal boxes, the correct route depending on the initial trial run, electrolytic lesions of the fimbria-fornix impaired performance (Walker and Olton, 1984). In addition to rodents, multi-unit mazes have been devised for monkeys, including a 3 x 3 matrix with 9 compartments (Traverse and Latto, 1986). The performance of this task was impaired after aspiration lesions of dorsolateral prefrontal or posterior parietal cortex (Traverse and Latto, 1986).

3. Hole-Board or Barnes Circular Maze

The hole-board task consists of finding one or more holes spread at the center of a wooden board to obtain food reward (Figure 6) (Oades and Isaacson, 1978). As with the tasks described above, when the goal remains the same, the test depends on reference memory. The acquisition of the hole-board task was slowed down after aspiration lesions of the hippocampus by increasing snout pokes into 4 empty holes among a matrix of 16 (Oades and Isaacson, 1978). The matrix can be multiplied to a cheese-board configuration by baiting one hole out of 177 (Kesner et al., 1989).

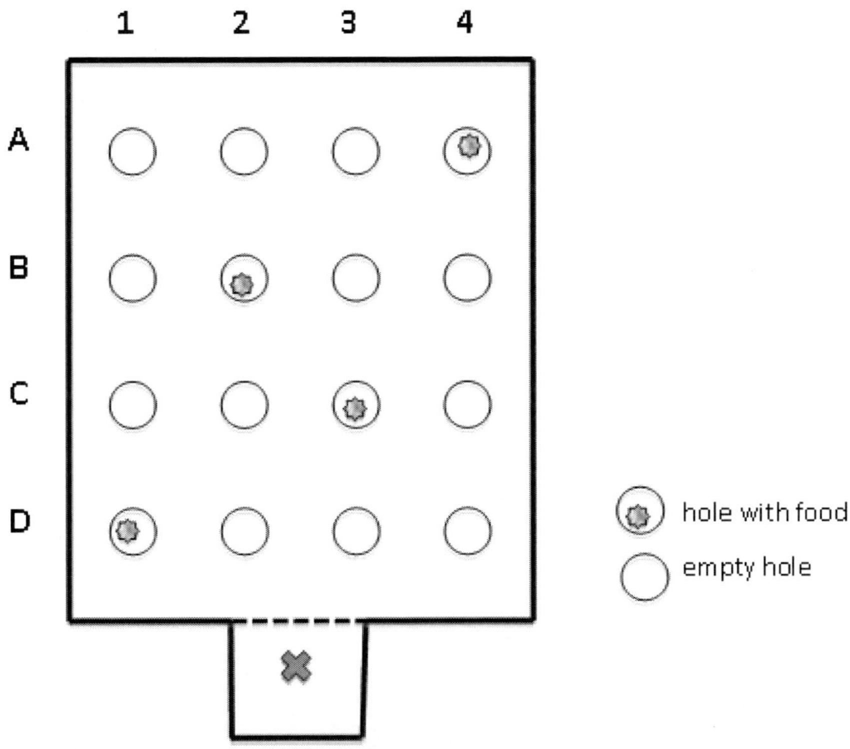

Figure 6. Diagram of a hole-board matrix motivated by food reward (Oades and Isaacson, 1978): the animal is placed at the start box (x) and must find food contained in holes A4, B2, C3, and D1.

The performance of the cheese-board task was impaired after aspiration lesions of posterior parietal but not medial prefrontal cortex (Kesner et al., 1989). In an adaptation of this test, a rat sees food in one of the holes and must return directly to it from seven starting points, performance being impaired after electrolytic lesions of the fimbria-fornix (Whishaw and Maaswinkel, 1998). In an allothetic version, a rat sees food in the middle of a table in the dark and returns to eat it in its home cage while the pathway is rotated, a task sensitive to aspiration lesions of the posterior cingulate cortex (Whishaw et al., 2001). A monkey version of the landmark task comprises a baited well chosen relative to a specific stimulus, a task sensitive to aspiration lesions of frontal eye field (Latto, 1986) or posterior parietal cortex (Mishkin et al., 1982), more particularly inferior parietal lobule area 7a (Latto, 1986).

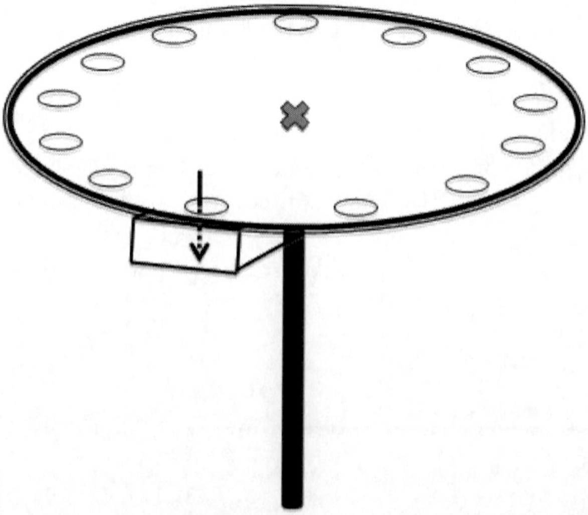

Figure 7. Diagram of the Barnes circular maze motivated by escape from bright lights (Barnes, 1979): the animal is placed at the center (x) of the apparatus and must find the one hole leading to escape (arrow).

As in the radial-arm maze (see below), the hole-board task offers a possibility of distinguishing reference from working memory (van der Staay et al., 1989). A reference memory error occurs whenever the animal enters a never-baited hole, whereas a working memory error occurs whenever the animal re-enters an empty but previously baited hole. Knife cuts into the fimbria-fornix impaired performance of both components (van der Staay et al., 1989). In a monkey version, aspiration lesions of sulcus principalis, including part of the frontal eye field, increased working memory errors (Passingham, 1985).

The Barnes circular maze consists of finding a hole at the periphery of the wooden board to escape from bright lights illuminating the apparatus (Figure 7) (Barnes, 1979; Barnes and McNaughton, 1985; O'Leary and Brown, 2013). Path length and latencies before reaching the escape hole as well as errors defined as nose-poking into empty holes are used as measures. The task is sensitive to traumatic impact of cortical tissue (Fox et al., 1999). There is some resemblance between hole-board configurations and a circular runway with eight wells providing water for thirsty rats (Figure 8), a task impaired by forceps crushing of the fimbria-fornix (O'Keefe et al., 1975).

4. LEFT-RIGHT DISCRIMINATION

In the left-right discrimination task, also dependent on reference memory, the animal must turn to the same side on every trial and thereby obtain a food reward or escape from footshocks or water in a T-maze (Figure 9), a Y-maze, or a Grice box, the latter consisting of a chamber with a partition separating right and left sides (Divac, 1971; Filali et al., 2009; Thompson, 1981).

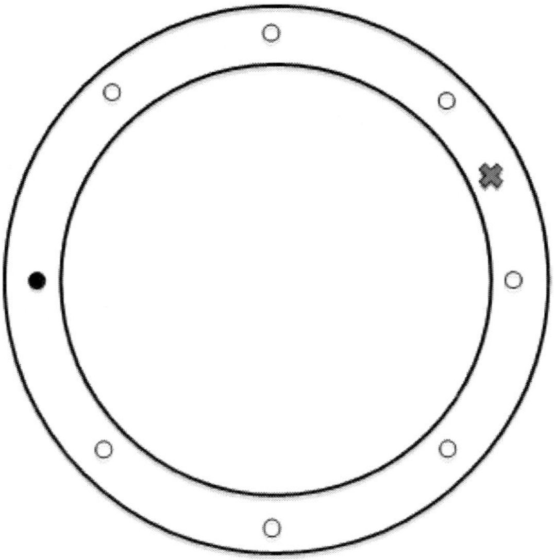

Figure 8. Diagram of a circular runway (O'Keefe et al., 1975): the animal (x) is placed in a runway and must find water contained in one (●) of eight wells (O).

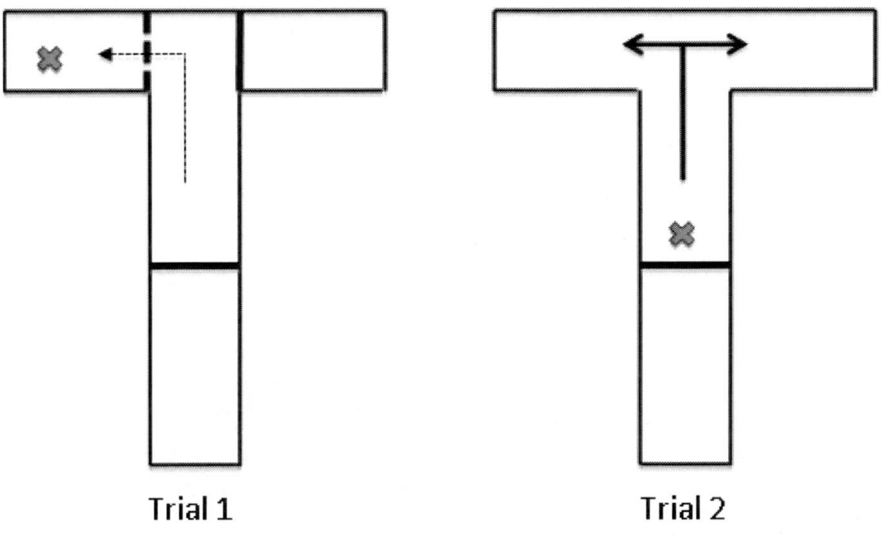

Figure 9. Diagram of a T-maze, used for left-right discrimination and alternation tests (Thompson, 1981). In the forced-choice spontaneous alternation test, the animal (x) is first placed in the stem and is forced to turn inside the left arm, exploring it for a specific amount of time. On trial 2, the choice trial, the animal may choose either to turn right (alternation) or left (perseveration). The stem may be shortened (bar division) to avoid excessively long movement times.

By using alley-ways between start and goal boxes, the animal need not be touched between trials (Shoji et al., 2012).

The acquisition of left-right discrimination in mice running for food reward was slowed down after excitotoxic lesions of posterior but not anterior cingulate cortex (Meunier et al.,

1991). Neither was the task affected by aspiration lesions of the medial frontal cortex in rats (Kolb et al., 1974), presumably because it does not require intricate step-by-step planning. With shock avoidance as motivation, the acquisition of left-right discrimination was slowed down after electrolytic lesions of the hippocampus (Thompson, 1981) or medial raphe (Thompson et al., 1989). Instead of 2 choices, the spatial discrimination task may be conducted with 4 choices in a cross-maze (Arns et al., 1999; Mayo et al., 1988; Ramos et al., 1998).

The acquisition of this version motivated by food reward was slowed down after excitotoxic lesions of the nucleus basalis magnocellularis in rats (Mayo et al., 1988) or hipppocampus in mice (Arns et al., 1999) and its performance impaired after electrolytic lesions of the dorsal hippocampus in rats (Ramos et al., 1998). A three-armed version has been adopted with the animal rewarded for entering the central arm among three, followed by a choice between the central one and one of the two non-rewarded side-arms (Guillou et al., 1993). The performance of this version was impaired after intrahippocampal injections of cysteamine in mice (Guillou et al., 1993). A reference memory version of the radial-arm maze has been devised whereby a single arm is rewarded among many (Liu and Bilkey, 1998). The acquisition of this version was slowed down after electrolytic lesions of the perirhinal cortex (Liu and Bilkey, 1998). Once criterion for the spatial discrimination is achieved, spatial reversals may be conducted whereby the rewarded side becomes non-rewarded and vice-versa (Divac, 1971). The acquisition (Kolb et al., 1974, 1983) and performance (Divac, 1971) of spatial reversals were impaired after aspiration lesions of medial frontal cortex in rats. Moreover, excitotoxic lesions of anterior or posterior cingulate cortex slowed down this measure in mice (Meunier et al., 1991). These results underline the importance of the prefrontal cortex in mental flexibility. The acquisition of spatial reversals was likewise slowed down after electrolytic lesions of dorsomedial thalamus, an area which projects to the medial frontal cortex (Kolb, 1977).

The same deficit occurred after electrolytic lesions of the septum (Schwartzbaum and Donovick, 1968) or excitotoxic lesions of medial septum plus nucleus basalis magnocellularis (Hepler et al., 1985; Peternel et al., 1988) or parabrachial nuclei (Kessler and Markowitsch, 1983). In addition, the performance of the task was impaired after razor blade cuts into the fimbria-fornix in rats (Greene and Stauff, 1974) or electrolytic lesions of the same area in monkeys (Mahut and Zola, 1973). In the shock avoidance paradigm, the acquisition of spatial reversals was delayed after electrolytic lesions of multiple brain regions, including hippocampus, medial raphe, and parietal cortex (Thompson, 1983; Thompson et al., 1989).

Instead of entry into maze arms, left-right discrimination and reversals may be investigated with bar-pressing (Béracochéa et al., 1989) or hole-poking (Lerer et al., 1985) responses. The acquisition of spatial reversals of hole-pokes was slowed down after excitotoxic lesions of the nucleus basalis magnocellularis (Lerer et al., 1985).

5. Object Displacement or Place Location

In the object displacement task, the animal explores two objects in a chamber, is taken out while one of the objects is displaced to another location, and then re-explores the objects (Thinus-Blanc et al., 1996). The learning measure used is the higher duration spent in exploring the displaced relative to the stationary object, a task dependent on the mouse hippocampus and posterior parietal cortex (Thinus-Blanc et al., 1996). This testing method has the advantage of requiring neither food nor water deprivation, neither water escape nor shock avoidance. The main disadvantage is that brain lesions sometimes influence initial object exploration, rendering interpretation of the retention test more difficult.

In another version of the test, hungry rats see two objects of different shape and are trained to head towards the object found in the same place irrespective of its shape (Becker et al., 1981). The performance of object place learning was deficient after electrolytic lesions of the fimbria-fornix as well as aspiration lesions of medial frontal cortex (Becker et al., 1981). An alternate version exists whereby a chamber replaces the object (Ennaceur and Meliani, 1992). In left-right conditional learning, monkeys can be trained to turn on one side (left or right) when placed on one corner (left or right) of a room and turn the opposite side when placed on the opposite corner (Gaffan and Harrison, 1989). In a conditional learning task dependent on color, success is achieved when the monkey turns left when a white tray is presented and turns right when an orange one is presented. Performance of either task was impaired after aspiration lesions of the fimbria-fornix.

6. Spatial Delayed Response

Spatial delayed response consists of giving the animal food on the left or right side of a panel and after a retention interval, placing another piece of food at the same place. Since the goal varies from one trial to the next (left or right), delayed response is conceived as a working memory task. The performance of the pre-operatively acquired task was impaired after aspiration lesions of the medial frontal cortex in rats (Kolb et al., 1974). Likewise, performances declined after prefrontal lobotomy (Rosvold et al., 1961) or aspiration lesions (Oscar-Berman, 1975; Pribram et al., 1952) of the dorsolateral prefrontal cortex in monkeys. The same deficit was obtained after combined aspiration lesions of hippocampus and amygdala, which send information to prefrontal cortex via thalamic nuclei (Zola-Morgan and Squire, 1985). Moreover, the acquisition of this task was slowed down after electrolytic lesions of two other brain regions connected with the prefrontal cortex, namely dorsomedial striatum (Battig et al., 1960) and dorsomedial thalamus (Isseroff et al., 1982).

7. Radial-Arm Maze

As with spatial delayed response, the radial-arm maze task requires spatial working memory. In the standard version of the test, hungry animals retrieve food pellets by entering once inside each of several maze arms (usually 6, 8, or 12) radiating from a central platform (Figure 10) (Walker and Olton, 1979). Maze arms may be added at the extremity of each arm to form a hierarchical pattern (Roberts, 1979). Unlike the Morris water maze task, the radial-arm maze task follows a win-shift foraging rule. Like the Morris water maze task, the radial-arm maze task depends on allocentric information provided by extra-maze cues, because the animals generally forage non-adjacent arms. Also akin to the Morris water maze task, the radial-arm maze task is more easily mastered by rats than mice (Mizumori et al., 1982). The maze is sometimes rotated from one trial to the next to prevent the use of olfactory cues.

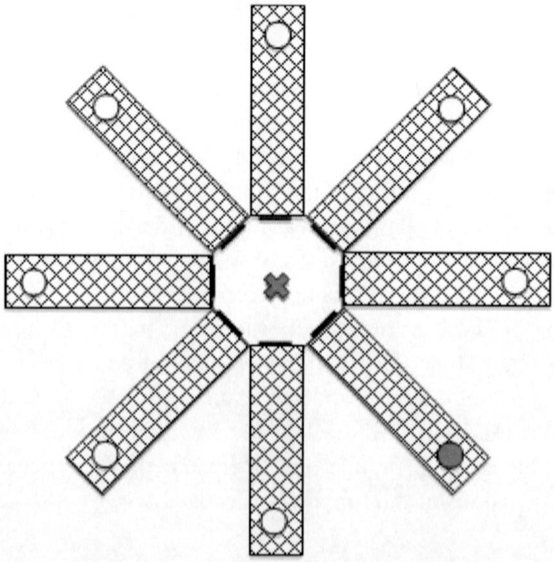

Figure 10. Diagram of a radial-arm maze (Walker and Olton, 1979): the animal (x) is placed at the center platform and must visit each maze arm once to obtain a food reward (●) while avoiding an arm that once contained food (O). An error is counted whenever the animal returns to an arm not containing food.

The task is dependent on visual cues, since aspiration lesions of primary and secondary visual cortex impaired performance (Foreman and Stevens, 1982; Goodale and Dale, 1981), as did electrolytic lesions of the superior colliculus, though to a lesser extent (Foreman and Stevens, 1982).

In this set-up, working memory can be dissociated from reference memory by baiting some arms and not others (Kesner et al., 1987; Olton and Papas, 1979). Re-entry into an empty but previously baited arm is defined as a working memory error and re-entry into a never-baited arm a reference memory error. Working memory is more labile than reference memory, as indicated by its accrued susceptibility to electroconvulsive shocks administered during the retention interval (Beatty et al., 1985).

7.1. Performance of the Pre-Operatively Acquired Task

In the radial-arm maze task, rats are usually pretrained to criterion and then tested postoperatively. Performance was impaired after damage to the fimbria-fornix (Walker and Olton, 1979), first described as increasing working but not reference memory errors (Olton and Papas, 1979). However, subsequent studies indicate that aspiration lesions of fimbria-fornix (Jarrard et al., 1984), entorhinal cortex (Jarrard et al., 1984), or hippocampus (Jarrard and Elmes, 1982) and electrolytic lesions of the hippocampus (Leis et al., 1984) increased both types of errors, as did kindling stimulation of the hippocampus (Lopes da Silva et al., 1986). Unlike these parts of the medial temporal lobe, damage to the amygdala was without any effect (Becker et al., 1980). Aspiration lesions of the medial prefrontal cortex also increased both types of errors, but aspiration lesions of the posterior parietal cortex only affected reference memory (Kesner et al., 1987). The task may be re-examined by distinguishing primary from recency components of memory. Damage to the dorsal hippocampus disrupted the primary but not the recency component, indicating an effect on long-term as opposed to short-term memory (Kesner and Novak, 1982).

In addition to the medial temporal lobe, performance of the radial-arm maze task depends on an intact diencephalon and midbrain, being impaired after electrolytic lesions of dorsomedial thalamus (Stokes and Best, 1988, 1990), mamillary bodies (Saravis et al., 1990; Sziklas and Petrides, 1993), or medial raphe (Wirtshafter and Asin, 1983). However, excitotoxic lesions of the first two regions had no effect (Béracochéa et al., 1989), a result which underlines the importance of fibers of passage.

7.2. Acquisition

When acquisition of the radial-arm maze task was assessed, errors increased after aspiration lesions of neocortical regions, namely medial frontal (Kolb et al., 1982, 1983), orbitofrontal (Kolb et al., 1983), or posterior parietal (Kolb and Walkey, 1987). Acquisition was also slowed down after electrolytic lesions of perirhinal cortex (Liu and Bilkey, 1998), fimbria-fornix (Packard et al., 1989), or medial raphe (Wirtshafter and Asin, 1983), and either electrolytic (Decker et al., 1992; Mitchell et al., 1982; Miyamoto et al., 1987) or excitotoxic (Decker et al., 1992) lesions of medial septum or nucleus basalis magnocellularis (Murray and Fibiger, 1985; Miyamoto et al., 1987).

As seen during performance of the task, acquisition appears to depend on the diencephalon, being delayed after electrolytic lesions of centromedial-parafascicularis or dorsomedial thalamus, though excitotoxic lesions of the former had no impact (M'Harzi et al., 1991b). In one study, the acquisition of working but not reference memory was delayed after excitotoxic lesions of the reticular thalamic nucleus (M'Harzi et al., 1991a), but in another, no such effect occurred (Wilton et al., 2001a).

7.3. Variations in Task Parameters

Radial-arm maze performance relies on intramaze or extramaze cues (Jarrard et al., 1984). When examined apart, aspiration lesions of the hippocampus or excitotoxic lesions of the subiculum increased errors in the place but not the cued task (Jarrard, 1983). The task has been adapted in an egocentric version by forcing the animal to forage adjacent maze arms (Kesner et al., 1989). The acquisition of this version was delayed after aspiration lesions of medial prefrontal but not posterior parietal cortex (Kolb et al., 1994). Aspiration lesions of medial prefrontal cortex also impaired performance (Kesner et al., 1989). The task has also been adapted to encourage a win-stay as opposed to the usual win-shift rule (DiMattia and Kesner, 1988a). Aspiration lesions of the posterior parietal cortex impaired performance of both win-shift and win-stay foraging rules (DiMattia and Kesner, 1988a). In yet another adaptation, win-stay arms may be signaled by a light source (Packard et al., 1989). Another variation is the wheel- or hexagonal-shaped maze (Figure 11) containing inner and outer alleys (Kessler et al., 1982). Performance of the hexagonal maze was impaired by excitotoxic lesions of the dorsomedial thalamus (Kessler et al., 1982).

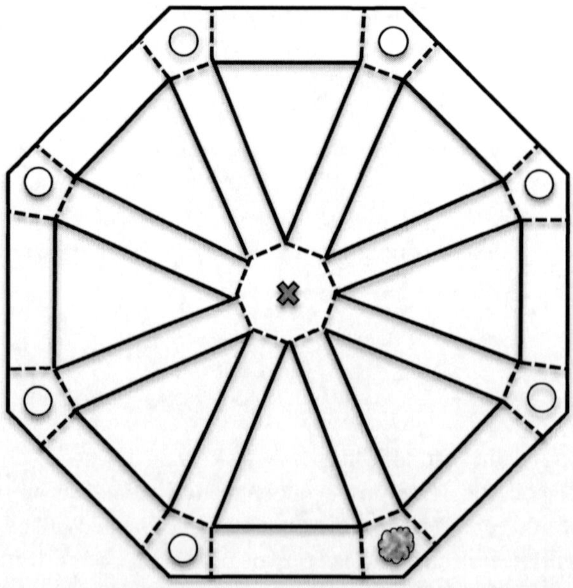

Figure 11. Diagram of a hexagonal-shaped maze (Kessler et al., 1982): the animal (x) is placed at the center platform and must visit each maze arm once to obtain a food reward (●) while avoiding an arm not containing food (O). The animal may circulate in alley-ways in addition to the arms. An error is counted whenever it heads in the wrong direction.

8. Left-Right Alternation and Non-Matching-to-Place

8.1. Spatial Alternation

The spatial alternation task (Figure 7) in T- or Y-mazes (Nonneman and Kolb 1979; Shoji et al., 2012) is a simplified version of the radial-arm maze task in which the number of choices is reduced to two (left or right) and the correct response is the reverse of spatial delayed response. It is often referred to as delayed reinforced alternation in view of the intertrial interval and the necessity of providing food reward or water escape to maximize performances, as opposed to the spontaneous alternation tendency (see below) motivated by the desire simply to explore the apparatus. Although counterintuitive to the human mind, rats acquire left-right alternations for food reward faster than left-right discriminations (Walker, 1984). The most likely reason is that survival of this species in the wild favors a win-shift as opposed to a win-stay strategy to avoid poisoning by human beings, the same phenomenon which underlies spontaneous alternation.

Reinforced alternation motivated by food reward is dependent on the rat limbic system, since its acquisition was slowed down after aspiration lesions of dorsomedial prefrontal cortex (Nonneman and Kolb 1979) or hippocampus (Means et al., 1971), electrolytic lesions of medial frontal (Thomas and Brito, 1980), prelimbic (Brito and Brito, 1990), or entorhinal (Ramirez and Stein, 1984) cortex, as well as posterodorsal septum (Brito and Brito, 1990; Brito and Thomas, 1981; Thomas and Brito, 1980) or mamillary bodies (Rosenstock et al., 1977), and after excitotoxic lesions of prelimbic cortex (Delatour and Gisquet-Verrier, 1996) or razor-blade cuts into fimbria-fornix (Greene and Stauff, 1974). The performance of the preoperatively acquired task was impaired after aspiration lesions of medial prefrontal cortex (Johnston et al., 1974; Larsen and Divac, 1978; Wikmark et al., 1973) or hippocampus (Racine and Kemble, 1965), electrolytic lesions of medial septum (Thomas et al., 1980) or posterodorsal septum (Brito and Thomas, 1981), and excitotoxic lesions of dorsomedial thalamus (Kessler and Markowitsch, 1981).

In addition to rats, the reinforced alternation task has been evaluated in cats, acquisition being slowed down after electrolytic lesions of the prefrontal cortex (Irle and Markowitsch, 1984). Performance of the task was impaired after excitotoxic lesions of dorsomedial thalamus (Markowitsch, 1982). It may be surprising to note that combined electrolytic lesions of seven limbic areas including anterior cingulate cortex and hippocampus had no effect on acquisition, presumably because of compensatory mechanisms following massive damage (Irle, 1985).

The task has been extended to monkeys, acquisition being slowed down after prefrontal lobotomy (Rosvold et al., 1961), aspiration lesions of dorsolateral prefrontal cortex (Pribram et al., 1952), and electrolytic lesions of caudate nucleus (Battig et al., 1960; Rosvold et al., 1958) or dorsomedial thalamus (Isseroff et al., 1982). Likewise, performance of the preoperatively acquired task was impaired after aspiration lesions of prefrontal cortex (Nixon and Passingham, 1999), electrolytic (Isseroff et al., 1982) or excitotoxic (Markowitsch, 1982) lesions of dorsomedial thalamus, and electrolytic lesions of mamillary bodies (Holmes et al., 1983) or dorsomedial thalamus (Butters and Rosvold, 1968), all these regions being interconnected in cortico-subcortical circuitries (chapter 3).

8.2. Non-Matching-to-Place

Closely related to delayed alternation is the non-matching-to-place task, whereby a sample run (left or right) is followed by a trial run to be executed on the opposite side (Granon et al., 1994). The parameters can be changed to its opposite, matching-to-place, in which rats must return to the same place as the trial run (Granon et al., 1994).

The acquisition of the non-matching-to-place principle was compromised by electrolytic lesions of several parts of the rat limbic system, including prelimbic cortex (Brito and Brito, 1990; Granon et al., 1994), medial septum (Rawlins and Olton, 1982), lateral septum (Rawlins and Olton, 1982), posterodorsal septum (Brito and Brito, 1990), fimbria-fornix (Rawlins and Olton, 1982; Warburton and Aggleton, 1999), mamillary bodies (Greene and Naranjo, 1986), or medial raphe (Asin and Fibiger, 1984), as well as excitotoxic lesions of fimbria-fornix (Aggleton et al., 1995), several thalamic nuclei including reticular (Collery et al., 1993), anterior (Aggleton et al., 1995; Collery et al., 1993; Warburton and Aggleton, 1999), or anterior dorsal/lateral dorsal (Wilton et al., 2001b), along with mamillary bodies (Aggleton et al., 1995) or medial raphe (Asin and Fibiger, 1984). The acquisition of the non-matching-to-place principle was also compromised by excitotoxic lesions of the hippocampus in mice (Pioli et al., 2014). In addition to non-matching-to-place, electrolytic lesions of the medial septum slowed down the acquisition of the matching-to-place principle in rats (Marighetto et al., 1998).

The rat limbic system is also involved in performance of the non-matching-to-place task, as indicated by impairments found after electrolytic lesions of prelimbic cortex (Granon et al.,

1994; Thomas and Spafford, 1984), posterodorsal septum (Thomas and Spafford, 1984), medial septum (Rawlins and Olton, 1982), fimbria-fornix (Aggleton et al., 1992; Neave et al., 1994), or dorsomedial thalamus (Hunt and Aggleton, 1991), aspiration lesions of hippocampus (Aggleton et al., 1986, 1992) or fimbria-fornix (Aggleton et al., 1991), and excitotoxic lesions of hippocampus (Kesslak and Gage, 1986), nucleus basalis magnocellularis (Salamone et al., 1984), or anterior thalamus (Aggleton et al., 1991), together with either electrolytic or excitotoxic lesions of medial septum or nucleus basalis magnocellularis (Helper et al., 1985). The impairment found after aspiration lesions of the fimbria-fornix was extended to monkeys (Murray et al., 1989).

The two arms may be continually changed but presented in pairs in plus- (Ulrich et al., 2014) or radial-arm (Saravis et al., 1990) maze configurations. Excitotoxic lesions of the anterior thalamus impaired performance of the task in a plus-maze configuration (Ulrich et al., 2014) and acquisition of the radial-arm maze configuration was slowed down after electrolytic lesions of the mamillary bodies (Saravis et al., 1990; Sziklas and Petrides, 1993).

8.3. Bar-Pressing Alternations

Instead of entries into maze arms, spatial alternation (Gross et al., 1965) and non-matching-to-place (Kirkby and Higgins, 1998) tasks have been analyzed by training rats to press levers. The acquisition and performance of alternating bar-presses were impaired after electrolytic lesions of hippocampus (Gross et al., 1965) or medial septum (Numan, 1991; Numan and Quaranta, 1990) in rats. In contrast, excitotoxic lesions of nucleus basalis magnocellularis slowed down acquisition but not performance (Dunnett et al., 1989). The acquisition of alternating bar-presses was slowed down after electrolytic lesions of the fimbria-fornix (Pisa, 1981), while performance was impaired after electrolytic lesions of medial frontal cortex (Dunnett et al., 1999; Van Haaren et al., 1985), fimbria-fornix (Neave et al., 1994), hippocampus (Wan et al., 1994), or dorsomedial thalamus (Gross et al., 1965), along with knife cuts into the perforant path (Kirkby and Higgins, 1998). A rarer version has been presented in which left-right spatial alternation is executed while pressing a lever inside one compartment and another lever inside a different compartment (Donovick, 1968), acquisition of this version being delayed after lesions of the anterior septum (Donovick, 1968).

9. Spontaneous Alternation

As with reinforced alternation, spontaneous alternation in a T- or Y-maze occurs when rats or mice shift their exploration of maze arms on successive trials, except that, in this paradigm, the animals are motivated only by their need to explore the maze and thereby satisfy their curiosity or to avoid returning to a familiar place and thereby avoid predation (Dember and Fowler, 1958; Lalonde, 2002). Though liable to be influenced by visual cues, spontaneous alternation is not dependent on them, as indicated by normal alternation rates found in rats blinded as a consequence of retinal lesions (Dember, 1959), a result which underlines the importance of kinesthetic cues. On the initial trial, both maze arms may be left open (free trial paradigm) or one of them may be closed (forced trial paradigm). On the second trial, spontaneous alternation rates are higher in the forced trial than the free trial paradigm, presumably because the former comprises a higher level of novelty since the barrier is removed (Dember and Fowler, 1959). Spontaneous alternation requires spatial or kinesthetic working memory in that the animal must remember where it had turned on the initial trial, but emotion-related factors are also relevant, since rates declined after rats received footshocks in the stem of the T-maze, causing an increase in anxiety (Rodriguez et al., 1992).

As with reinforced alternation, spontaneous alternation depends on proper functioning of the rodent limbic system. Indeed, spontaneous alternation rates declined after aspiration lesions of medial prefrontal cortex (Divac et al., 1975), either aspiration (Means et al., 1971) or electrolytic (Dalland, 1970, 1976; Ellen and Deloache, 1968; Isseroff, 1979; Johnson et al., 1977; Roberts et al., 1962; Stevens and Cowey, 1973) lesions of hippocampus, and after electrolytic lesions of septum (Dalland, 1970; Douglas and Raphelson, 1966; Thomas, 1972), medial septum (Clody and Carlton, 1969), dorsomedial thalamus (Weis and Means, 1980), or medial raphe (Asin and Fibiger, 1984) in rats. Likewise, rates declined after excitotoxic lesions of prelimbic cortex (Delatour and Gisquet-Verrier, 1996), anterior thalamus (Bérachochéa and Jaffard, 1994), or mamillary bodies (Béracochéa and Jaffard, 1987) in rats and electrolytic lesions of mamillary bodies in mice (Béracochéa and Jaffard, 1987; Tako et al., 1988).

Instead of alternation rates, time spent exploring the familiar arm of the T-maze may be used as an alternative measure (M'Harzi et al., 1991b). Rats with electrolytic but not excitotoxic lesions of centromedial-parafascicularis thalamus spent more time in the familiar arm, indicating a decreased tendency to alternate, whereas those with electrolytic lesions of the dorsomedial thalamus did not (M'Harzi et al., 1991b). In contrast, deficits on this measure were revealed after excitotoxic lesions of either reticular or anterior thalamus (Collery et al., 1993).

REFERENCES

Aggleton, JP; Hunt, PR; Rawlins, JNP. The effects of hippocampal lesions upon spatial and non-spatial tests of working memory. *Behav. Brain Res.,* 1986, 19, 133-46.

Aggleton, JP; Keith, AB; Sahgal, A. Both fornix and anterior thalamic, but not mammillary, lesions disrupt delayed non-matching-to-position memory in rats. *Behav. Brain Res.,* 1991, 44, 151-61.

Aggleton, JP; Neave, N; Nagle, S; Hunt, PR. A comparison of the effects of anterior thalamic, mamillary body and fornix lesions on reinforced spatial alternation. *Behav. Brain Res.,* 1995, 68, 91-101.

Aggleton, JP; Keith, AB; Rawlins, JN; Hunt, PR; Sahgal, A. Removal of the hippocampus and transection of the fornix produce comparable deficits on delayed non-matching to position by rats. *Behav. Brain Res.,* 1992, 52, 61-71.

Ain, BR; Lubar, JF; Moon, RD; Kulig, BM. Effect of septal and neocortical damage on complex maze learning. *Physiol. Behav.,* 1969, 4, 235-238.

Arns, M; Sauvage, M; Steckler, T. Excitotoxic hippocampal lesions disrupt allocentric spatial learning in mice: effects of strain and task demands. *Behav. Brain Res.,* 1999, 106, 151-64.

Asin, KE; Fibiger, HC. Spontaneous and delayed spatial alternation following damage to specific neuronal elements within the nucleus medianus raphe. *Behav. Brain Res.,* 1984, 13, 241-50.

Bannerman, DM; Yee, BK; Good, MA; Heupel, MJ; Iversen, SD; Rawlins, JN. Double dissociation of function within the hippocampus: a comparison of dorsal, ventral, and complete hippocampal cytotoxic lesions. *Behav. Neurosci.,* 1999, 113, 1170-88.

Barnes, CA. Memory deficits associated with senescence: a neurophysiological and behavioral study in the rat. *J. Comp. Physiol. Psychol.,* 1979, 93, 74-104.

Barnes, CA; McNaughton, BL. Spatial information: how and where is it stored? In: NM Weinberger, JL McGaugh, G Lynch (eds) *Memory systems of the brain*. New York: Guildford Press, pp 49-61, 1985.

Barone, S Jr; Nanry, KP; Mundy, WR; McGinty, JF; Tilson, HA. Spatial learning deficits are not solely due to cholinergic deficits following medial septal lesions with colchicine. *Psychobiology,* 1991, 19, 41-50.

Battig, K; Rolsvold, HE; Mishkin, M. Comparison of the effects of frontal and caudate lesions on delayed response and alternation in monkeys. *J. Comp. Physiol. Psychol.,* 1960, 53, 400-4.

Baxter, MG; Bucci, DJ; Gorman, LK; Wiley, RG; Gallagher, M. Selective immunotoxic lesions of basal forebrain cholinergic cells: effects on learning and memory in rats. *Behav. Neurosci.,* 1995, 109, 714-22.

Baxter, MG; Bucci, DJ; Sobel, TJ; Williams, MJ; Gorman, LK; Gallagher, M. Intact spatial learning following lesions of basal forebrain cholinergic neurons. *NeuroReport,* 1996, 7, 1417-20.

Bear, MF; Connors BW. Paradiso MA. *Neuroscience: exploring the brain*. Baltimore: Williams & Wilkins, pp 514-45, 1996.

Beatty, WW; Bierley, RA; Rush, JR. Spatial memory in rats: electroconvulsive shock selectively disrupts working memory but spares reference memory. *Behav. Neural Biol.,* 1985, 44, 403-414.

Becker, JT; Walker, JA; Olton, DS. Neuroanatomical bases of spatial memory. *Brain Res.,* 1980, 200, 307-20.

Becker, JT; Olton, DS; Anderson, CA; Breitinger, ER. Cognitive mapping in rats: the role of the hippocampal and frontal system in retention and reversal. *Behav. Brain Res.,* 1981, 3, 1-22.

Beninger, RJ; Kühnemann, S; Ingles, JL; Jhamandas, K; Boegman, RJ. Mnemonic deficits in the double Y-maze are related to the effects of nucleus basalis injections of ibotenic and quisqualic acid on choline acetyltransferase in the rat amygdala. *Brain Res. Bull.,* 1994, 35, 147-52.

Béracochéa, DJ; Jaffard, R. Impairment of spontaneous alternation behavior in sequential test procedures following mammillary body lesions in mice: evidence for time-dependent interference-related memory deficits. *Behav. Neurosci.,* 1987, 101, 187-197.

Béracochéa, DJ; Jaffard, R; Jarrard, LE. Effects of anterior or dorsomedial thalamic ibotenic lesions on learning and memory in rats. *Behav. Neural Biol.,* 1989, 51, 364-76.

Béracochéa, DJ; Jaffard, R. Effects of anterior thalamic lesions on spatial memory in mice. *NeuroReport,* 1994, 5, 917-20.

Berger-Sweeney, J; Stearns, NA; Murg, SL; Floerke-Nashner, LR; Lappi, DA; Baxter, MG. Selective immunolesions of cholinergic neurons in mice: effects on neuroanatomy, neurochemistry, and behavior. *J. Neurosci.,* 2001, 21, 8164-73.

Bonner, JT. *The evolution of culture in animals*. Princeton: Princeton University Press, p 115, 1980.

Boyd, MG; Thomas, RK. Posterior association cortex lesion in rats: mazes patterns discrimination, and reversal learning. *Physiol. Psychol.,* 1977, 5, 455-61.

Brandeis, R; Brandys, Y; Yehuda, S. The use of the Morris Water Maze in the study of memory and learning. *Int. J. Neurosci.,* 1989, 48, 29-69.

Brito, GNO; Brito, SO. Septohippocampal system and the prelimbic sector of frontal cortex: a neuropsychological battery analysis in the rat. *Behav. Brain Res.,* 1990, 36, 127-46.

Brito, GNO; Thomas, GJ. T-maze alternation, response patterning, and septo-hippocampal circuitry in rats. *Behav. Brain Res.,* 1981, 3, 319-40.

Brun, VH; Otnaess, MK; Molden, S; Steffenach, HA; Witter, MP; Moser, MB; Moser, EI. Place cells and place recognition maintained by direct entorhinal-hippocampal circuitry. *Science,* 2002, 296, 2243-6.

Burwell, RD; Saddoris, MP; Bucci, DJ; Wiig, KA. Corticohippocampal contributions to spatial and contextual learning. *J. Neurosci.,* 2004, 24, 3826-36.

Butters, N; Rosvold, HE. Effect of caudate and septal nuclei lesions on resistance to extinction and delayed-alternation. *J. Comp. Physiol. Psychol.,* 1968, 65, 397-403.

Buzsaki, G; Bors, L; Nagy, F; Eidelberg, E. Spatial mapping, working memory, and the fimbria-fornix system. *J. Comp. Physiol. Psychol.,* 1982, 96, 26-34.

Cain, DP; Saucier, D; Hargreaves, EL; Wilson, E; DeSouza, JF. Polypropylene pellets as an inexpensive reusable substitute for milk in the Morris milk maze. *J. Neurosci. Methods,* 1993, 49, 193-7.

Churchwell, JC; Morris, AM; Musso, ND; Kesner, RP. Prefrontal and hippocampal contributions to encoding and retrieval of spatial memory. *Neurobiol. Learn. Mem.,* 2010, 93, 415-21.

Clody, DE; Carlton, PL. Behavioral effects of lesions of the medial septum of rats. *J. Comp. Physiol. Psychol.,* 1969, 67, 344-51.

Collery, M; M'Harzi, M; Delacour, J. Lesions of reticularis thalamic nucleus impair spatial working memory in rats. *Neurosci. Res. Comm.,* 1993, 12, 41-9.

Dalland, T. Response perseveration of rats with dorsal hippocampal lesions. *Behav. Biol.,* 1976, 17, 473-84.

Dalland, T. Response and stimulus perseveration in rats with septal and dorsal hippocampal lesions. *J. Comp. Physiol. Psychol.,* 1970, 71, 114-8.

de Bruin, JP; Moita, MP; de Brabander, HM; Joosten, RN. Place and response learning of rats in a Morris water maze: differential effects of fimbria fornix and medial prefrontal cortex lesions. *Neurobiol. Learn. Mem.,* 2001, 75, 164-78.

de Bruin, JP; Sànchez-Santed, F; Heinsbroek, RP; Donker, A; Postmes, P. A behavioural analysis of rats with damage to the medial prefrontal cortex using the Morris water maze: evidence for behavioural flexibility, but not for impaired spatial navigation. *Brain Res.,* 1994, 652, 323-33.

Decker, MW; Curzon, P; Brioni, JD. Influence of separate and combined septal and amygdala lesions on memory, acoustic startle, anxiety, and locomotor activity in rats. *Neurobiol. Learn. Mem.,* 1995, 64, 156-68.

Decker, MW; Radek, RJ; Majchrzak, MJ; Anderson, DJ. Differential effects of medial septal lesions on spatial-memory tasks. *Psychobiology,* 1992, 20, 9-17.

Delatour, B; Gisquet-Verrier, P. Prelimbic cortex specific lesions disrupt delayed-variable response tasks in the rat. *Behav. Neurosci.,* 1996, 110, 1282-98.

Dember, WN. Spontaneous alternation in peripherally blinded rats. *Can. J. Psychol.,* 1959, 12, 219-21.

Dember, WN; Fowler, H. Spontaneous alternation behavior. *Psychol. Bull.,* 1958, 55, 412-28.

Dember, WN; Fowler, H. Spontaneous alternation after free and forced trials. *Can. J. Psychol.,* 1959, 13, 151-4.

D'Hooge, R; De Deyn, PP. Applications of the Morris water maze in the study of learning and memory. *Brain Res. Rev.,* 2001, 36, 60-90.

DiMattia, BV; Kesner, RP. Role of the posterior parietal association cortex in the processing of spatial event information. *Behav. Neurosci.,* 1988a, 102, 397-403.

DiMattia, BV; Kesner, RP. Spatial cognitive maps: differential role of parietal cortex and hippocampal formation. *Behav. Neurosci.,* 1988b, 102, 471-80.

Divac, I. Frontal lobe system and spatial reversal in the rat. *Neuropsychologia,* 1971, 9, 175-83.

Divac, I; Wiskmar, RGE; Gade, A. Spontaneous alternation in rats with lesions of the frontal lobes: an extension of the frontal lobe syndrome. *Physiol. Psychol.,* 1975, 3, 39-42.

Dolleman-van der Weel, MJ; Morris, RG; Witter, MP. Neurotoxic lesions of the thalamic reuniens or mediodorsal nucleus in rats affect non-mnemonic aspects of watermaze learning. *Brain Struct. Funct.,* 2009, 213, 329-42.

Donovick, PJ. Effects of localized septal lesions on hippocampal EEG activity and behavior in rats. *J. Comp. Physiol. Psychol.,* 1968, 66, 569-78.

Douglas, RJ; Raphelson, AC. Spontaneous alternation and septal lesions. *J. Comp. Physiol. Psychol.,* 1966, 62, 320-2.

Dunnett, SB; Nathwani, F; Brasted, PJ. Medial prefrontal and neostriatal lesions disrupt performance in an operant delayed alternation task in rats. *Behav. Brain Res.,* 1999, 106, 13-28.

Dunnett, SB; Rogers, DC; Jones, GH. Effects of nucleus basalis magnocellularis lesions in rats on delayed matching and non-matching to position tasks. *Eur. J. Neurosci.,* 1989, 1, 395-406.

Eichenbaum, H; Stewart, C; Morris, RGM. Hippocampal representation in place learning. *J. Neurosci.,* 1990, 10, 3531-42.

Ellen, P; Deloache, J. Hippocampal lesions and spontaneous alternation behaviour in the rat. *Physiol. Behav.,* 1968, 3, 857-60.

Emlen, ST. The stellar-orientation system of a migratory bird. *Sci. Am.,* 1975, 233, 102-11.

Ennaceur, A; Meliani, K. A new one-trial test for neurobiological studies of memory in rats. III. Spatial vs. non-spatial working memory. *Behav. Brain Res.,* 1992, 51, 83-92.

Espinoza, S; Pinto-Hamuy, T; Passig, C; Carreño, F; Marchant, F; Urzúa, C. Deficit in the water-maze after lesions in the anteromedial extrastriate cortex in rats. *Physiol. Behav.,* 1999, 66, 493-6.

Feigenbaum, JD; Rolls, ET. Allocentric and egocentric spatial information processing in the hippocampal formation of the behaving primate. *Psychobiology,* 1991, 19, 21-40.

Filali, M; Lalonde, R; Rivest, S. Cognitive and non-cognitive behaviors in an APPswe/PS1 bigenic model of Alzheimer's disease. *Genes Brain Behav.,* 2009, 8, 143-8.

Foreman, N; Stevens, R. Visual lesions and radial maze performance in rats. *Behav. Neural Biol,* 1982, 36, 126-36.

Foster, TC; Castro, CA; McNaughton, BL. Spatial selectivity of hippocampal neurons: dependence on preparedness for movement. *Science,* 1989, 244, 1580-2.

Fox, GB; LeVasseur, RA; Faden, AI. Behavioral responses of C57BL/6, FVB/N, and 129/SvEMS mouse strains to traumatic brain injury: implications for gene targeting approaches to neurotrauma. *J. Neurotrauma,* 1999, 16, 377-89.

Gaffan, D; Harrison, S. Place memory and scene memory: effects of fornix transection in the monkey. *Exp. Brain Res.,* 1989, 74, 202-12.

Goodale, MA; Dale, RHI. Radial-maze performance in the rat following lesions of posterior neocortex. *Behav. Brain Res.,* 1981, 3, 273-88.

Granon, S; Vidal, C; Thinus-Blanc, C; Changeux, JP; Poucet, B. Working memory, response selection, and effortful processing in rats with medial prefrontal lesions. *Behav. Neurosci.,* 1994, 108, 883-91.

Greene, E; Naranjo, JN. Thalamic role in spatial memory. *Behav. Brain Res.,* 1986, 19, 123-31.

Greene, E; Stauff, C. Behavioral role of hippocampal connections. *Exp. Neurol.,* 1974. 45, 141-60.

Gross, CG; Chorover, SL; Cohen, SM. Caudate, cortical, hippocampal and dorsal thalamic lesions in rats: alternation and Hebb-Williams maze performance. *Neuropsychologia,* 1965, 3, 53-68.

Guillou, J-L; Micheau, J; Jaffard, R. Effects of intrahippocampal injections of somatostatin and cysteamine on spatial discrimination learning in mice. *Psychobiology,* 1993, 21, 265-71.

Hagan, JJ; Salamone, JD; Simpson, J; Iversen, SD; Morris, RG. Place navigation in rats is impaired by lesions of medial septum and diagonal band but not nucleus basalis magnocellularis. *Behav. Brain Res.*, 1988, 27, 9-20.

Hardman, R; Evans, DJ; Fellows, L; Hayes, B; Rupniak, HT; Barnes, JC; Higgins, GA. Evidence for recovery of spatial learning following entorhinal cortex lesions in mice. *Brain Res.,* 1997, 758, 187-200.

Hebb, DO; Williams, K. A method of rating animal intelligence. *J. Genet. Psychol.,* 1946, 34, 59-65.

Hepler, DJ; Olton, DS; Wenk, G; Coyle, JT. Lesions in nucleus basalis magnocellularis and medial septal area of rats produce qualitatively similar memory impairments. *J. Neurosci.,* 1985, 5, 866-73.

Herrmann, T; Poucet, B; Ellen, P. Spatial problem solving in the rat following medial frontal lesions. *Physiol. Psychol.,* 1985a, 13, 21-5.

Herrmann, T; Poucet, B; Ellen, P. Spatial problem solving in a dual runway task by normal and septal rats. *Behav. Neurosci.,* 1985b, 99, 631-7.

Hoh, TE; Kolb, B; Eppel, A; Vanderwolf, CH; Cain, DP. Role of the neocortex in the water maze task in the rat: a detailed behavioral and Golgi-Cox analysis. *Behav. Brain Res.,* 2003, 138, 81-94.

Holmes, EJ; Jacobson, S; Stein, BM; Butters, N. Ablations of the mammillary nuclei in monkeys: effects on postoperative memory. *Exp. Neurol.,* 1983, 81, 97-113.

Hoplight, BJ II; Boehm, GW; Hyde, LA; Deni, R; Denenberg, VH. A computer-aided procedure for measuring Hebb-Williams maze performance. *Physiol. Behav.,* 1996, 60, 1171-6.

Hunt, PR; Aggleton, JP. Medial dorsal thalamic lesions and working memory in the rat. *Behav. Neural Biol.,* 1991, 55, 227-46.

Irle, E. Combined lesions of septum, amygdala, hippocampus, anterior thalamus, mamillary bodies and cingulate and subicular cortex fail to impair the acquisition of complex learning tasks. *Exp. Brain Res.,* 1985, 58, 346-61.

Irle, E; Markowitsch, HJ. Differential effects of prefrontal lesions and combined prefrontal and limbic lesions on subsequent learning performance in the cat. *Behav. Neurosci.,* 1984, 98, 884-97.

Isseroff, A. Limited recovery of spontaneous alternation after extensive hippocampal damage: evidence for a memory impairment. *Exp. Neurol.,* 1979, 64, 284-94.

Isseroff, A; Rolvold, HE; Galkin, TW; Goldman-Rakic, PS. Spatial memory impairments following damage to the mediodorsal nucleus of the thalamus in rhesus monkeys. *Brain Res.,* 1982, 232, 97-113.

Iivonen, H; Nurminen, L; Harri, M; Tanila, H; Puoliväli, J. Hypothermia in mice tested in Morris water maze. *Behav. Brain Res.,* 2003, 141, 207-13.

Jackson, WJ; Strong, PN Jr. Differential effects of hippocampal lesions upon sequential tasks and maze learning by the rat. *J. Comp. Physiol. Psychol.,* 1969, 68, 442-50.

Jankowski, MM; Ronnqvist, KC; Tsanov, M; Vann, SD; Wright, NF; Erichsen, JT; Aggleton, JP; O'Mara, SM. The anterior thalamus provides a subcortical circuit supporting memory and spatial navigation. *Front. Syst. Neurosci.,* 2013, 7, 45.

Jarrard, LE. Selective hippocampal lesions and behavior: effects of kainic acid lesions on performance of place and cue tasks. *Behav. Neurosci.,* 1983, 97, 873-89.

Jarrard, LE; Okaichi, H; Steward, O; Goldschmidt, RB. On the role of hippocampal connections in the performance of place and cue tasks: comparisons with damage to hippocampus. *Behav. Neurosci.,* 1984, 98, 946-54.

Johnson, CT; Olton, DS; Gage, F III; Jenko, PG. Damage to hippocampus and hippocampal connections: effects on DRL and spontaneous alternation. *J. Comp. Physiol. Psychol.,* 1977, 91, 508-22.

Johnston, VS; Hart, M; Howell W. The nature of the medial wall deficit in the rat. *Neuropsychologia,* 1974, 12, 497-503.

Kametani, H; Spangler, EL; Bresnahan, EL; Kobayashi; S, Long, JM; Ingram, DK. Impaired acquisition in a 14-unit T-maze following medial septal lesions in rats is correlated with lesion size and hippocampal acetylcholinesterase staining. *Physiol. Behav.,* 1993, 53, 221-8.

Kelsey, JE; Landry BA. Medial septal lesions disrupt spatial mapping ability in rats. *Behav. Neurosci.,* 1988, 102, 289-93.

Kesner, RP; Novak, JM. Serial position curve in rats: role of the dorsal hippocampus. *Science,* 1982, 218, 173-75.

Kesner, RP; DiMattia, BV; Crutcher, KA. Evidence for neocortical involvement in reference memory. *Behav. Neural Biol.,* 1987, 47, 40-53.

Kesner, RP; Farnsworth, G; DiMattia, BV. Double dissociation of egocentric and allocentric space following medial prefrontal and parietal cortex lesions in the rat. *Behav. Neurosci.,* 1989, 103, 956-61.

Kesner, RP; Farnsworth, G; Kametani, H. Role of parietal cortex and hippocampus in representing spatial information. *Cereb. Cortex,* 1991, 1, 367-73.

Kesslak, JP; Gage, FH III. Recovery of spatial alternation deficits following selective hippocampal destruction with kainic acid. *Behav. Neurosci.,* 1986, 100, 280-3.

Kessler, J; Markowitsch, HJ. Delayed-alternation performance after kainic acid lesions of the thalamic mediodorsal nucleus and the ventral tegmental area in the rat. *Behav. Brain Res.,* 1981, 3, 125-30.

Kessler, J; Markowitsch, HJ. Perseverative effects of parabrachial lesions: a mechanism of reduced activation? *Behav. Neural Biol.,* 1983, 39, 40-51.

Kessler, J; Markowitsch, HJ; Otto, B. Subtle but distinct impairments of rats with chemical lesions in the thalamic mediodorsal nucleus, tested in a radial arm maze. *J. Comp. Physiol. Psychol.,* 1982, 96, 712-20.

Kimble, DP. The effects of bilateral hippocampal lesions in rats. *J. Comp. Physiol. Psychol.,* 1963, 56, 272-83.

Kimble, DP; Jordan, WP; BreMiller, R. Further evidence for latent learning in hippocampal-lesioned rats. *Physiol. Behav.,* 1982, 29, 401-7.

Kirkby, DL; Higgins, GA. Characterization of perforant path lesions in rodent models of memory and attention. *Eur. J. Neurosci.,* 1998, 10, 823-38.

Kolb, B. Studies on the caudate-putamen and the dorsomedial thalamic nucleus of the rat: implications for mammalian frontal-lobe functions. *Physiol. Behav.,* 1977, 18, 237-44.

Kolb, B. Functions of frontal cortex of the rat: a comparative review. *Brain Res. Rev.,* 1984, 8, 65-98.

Kolb, B; Walkley, J. Behavioural and anatomical studies of the posterior parietal cortex in the rat. *Behav. Brain Res.,* 1987, 23, 127-45.

Kolb, B; Buhrmann, K; McDonald, R; Sutherland, RJ. Dissociation of the medial prefrontal, posterior parietal, and posterior temporal cortex for spatial navigation and recognition memory in the rat. *Cerebr. Cortex,* 1994, 6, 664-80.

Kolb, B; Nonneman, AJ; Singh, RK. Double dissociation of spatial impairments and perseveration following selective prefrontal lesions in rats. *J. Comp. Physiol. Psychol.,* 1974, 87, 772-80.

Kolb, B; Pittman K; Sutherland, RJ; Whishaw, IQ. Dissociation of the contributions of the prefrontal cortex and dorsomedial thalamic nucleus to spatially guided behavior in the rat. *Behav. Brain Res.,* 1982, 6, 365-78.

Kolb, B; Sutherland, RJ; Whishaw, IQ. A comparison of the contributions of the frontal and parietal association cortex to spatial localization in rats. *Behav. Neurosci.,* 1983, 97, 13-27.

Lalonde, R. The neurobiological basis of spontaneous alternation. *Neurosci. Biobehav. Rev.,* 2002, 26, 91-104.

Lalonde, R; Filali, M; Bensoula, AN; Monnier, C; Guastavino, J-M. Spatial learning in a Z-maze by cerebellar mutant mice. *Physiol. Behav.,* 1996, 59, 83-6.

Larsen, JK; Divac, I. Selective ablations within the prefrontal cortex of the rat and performance of delayed alternation. *Physiol. Psychol.,* 1978, 6, 15-7.

Lashley, KS. *Brain mechanisms and intelligence*. Chicago: University of Chicago Press, 1929.

Latto, R. The role of inferior parietal cortex and the frontal eye-fields in visuospatial discriminations in the macaque monkey. *Behav. Brain Res.,* 1986, 22, 41-52.

Leis, T; Pallage, V; Toniolo, G; Will, B. Working memory theory of hippocampal function needs qualification. *Behav. Neural Biol.,* 1984, 42, 140-57.

Lerer, B; Warner, J; Friedman, E; Vincent, G; Gamzu, E. Cortical cholinergic impairment and behavioral deficits produced by kainic acid lesions of rat magnocellular basal forebrain. *Behav. Neurosci.,* 1985, 99, 661-77.

Lines, CR; Milner, AD. A deficit in ambient visual guidance following superior colliculus lesions in rats. *Behav. Neurosci.,* 1985, 99, 707-16.

Liu, P; Bilkey, DK. Lesions of perirhinal cortex produce spatial memory deficits in the radial maze. *Hippocampus,* 1998, 8, 114-21.

Lopes da Silva, FH; Gorter, JA; Wadman, WJ. Kindling of the hippocampus induces spatial memory deficits in the rat. *Neurosci. Lett.,* 1986, 63, 115-20.

Mahut, H; Zola, SM. A non-modality specific impairment in spatial learning after fornix lesions in monkeys. *Neuropsychologia,* 1973, 11, 255-69.

Marighetto, A; Yee, BK; Rawlins, JNP. The effects of cytotoxic entorhinal lesions and electrolytic medial septal lesions on the acquisition and retention of a spatial working memory task. *Exp. Brain Res.,* 1998, 119, 517-28.

Markowitsch, HJ. Retention performance of a learned delayed-alternation task after chemical lesions of the cats mediodorsal nucleus. *Behav. Brain Res.,* 1982, 4, 263-77.

Marston, HM; Everitt, BJ; Robbins, TW. Comparative effects of excitotoxic lesions of the hippocampus and septum/diagonal band on conditional visual discrimination and spatial learning. *Neuropsychologia,* 1993, 31, 1099-118.

Mason, KI; Mallet, PE; Jhamandas, K; Boegman, RJ; Beninger, RJ. Nucleus basalis injections of N-methyl-D-aspartate enhance memory of rats in the double Y-maze. *Brain Res. Bull.,* 1999, 48, 65-71.

Mayo, W; Kharouby, M; Le Moal, M; Simon, H. Memory disturbances following ibotenic acid injections in the nucleus basalis magnocellularis of the rat. *Brain Res.,* 1988, 455, 213-22.

McNamara, RK; Skelton, RW. The neuropharmacological and neurochemical basis of place learning in the Morris water maze. *Brain Res. Rev.,* 1993, 18, 33-49.

Means, LW. Rats acquire win-stay more readily than win-shift in a water escape situation. *Anim. Learn. Behav.*, 1988, 16, 303-11.

Means, LW; Leander, JD; Isaacson, RL. The effects of hippocampectomy on alternation behavior and response to novelty. *Physiol. Behav.,* 1971, 6, 17-22.

Meunier, M; Jaffard, R; Destrade, C. Differential involvement of anterior and posterior cingulate cortices in spatial discriminative learning in a T-maze in mice. *Behav. Brain Res.,* 1991, 44, 133-43.

Meunier, M; Saint-Marc, M; Destrade, C. The Hebb-Williams test to assess recovery of learning after limbic lesions in mice. *Physiol. Behav.,* 1986, 37, 909-13.

M'Harzi, M; Collery, M; Delacour, J. First evidence of a possible role of the reticular thalamic nucleus in working memory in rats. *Neurosci. Res. Comm.,* 1991a, 8, 167-74.

M'Harzi M; Jarrard, LE; Willig, F; Palacios, A; Delacour, J. Selective fimbria and thalamic lesions differentially impair forms of working memory in rats. *Behav. Neural Biol.,* 1991b, 56, 221-30.

Mishkin M; Lewis, ME; Ungerleider, LG. Equivalence of parieto-preoccipital subeareas for visuaospatial ability in monkeys. *Behav. Brain Res.,* 1982, 6, 41-55.

Mitchell, SJ; Rawlins, JNP; Steward, O; Olton, DS. Medial septal area lesions disrupt theta rhythm and cholinergic staining in medial entorhinal cortex and produce impaired radial arm maze behavior in rats. *J. Neurosci.*, 1982, 2, 292-302.

Miyamoto, M; Kato, J; Narumi, S; Nagaoka, A. Characteristics of memory impairment following lesioning of the basal forebrain and medial septal nucleus in rats. *Brain Res.,* 1987, 419, 19-31.

Mizumori, SJY; Rosenzweig, MR; Kermisch, MG. Failure of mice to demonstrate spatial memory in the radial maze. *Behav. Neural Biol.,* 1982, 35, 33-45.

Moreau, PH; Cosquer, B; Jeltsch, H; Cassel, JC; Mathis, C. Neuroanatomical and behavioral effects of a novel version of the cholinergic immunotoxin mu p75-saporin in mice. *Hippocampus,* 2008, 18, 610-22.

Morris, RG; Schenk, F; Tweedie, F; Jarrard, LE. Ibotenate lesions of hippocampus and/or subiculum: dissociating components of allocentric spatial learning. *Eur. J. Neurosci.,* 1990, 2, 1016-28.

Morris, RGM; Garrud, P; Rawlins, JNP; O'Keefe, J. Place navigation impaired in rats with hippocampal lesions. *Nature,* 1982, 292, 681-3.

Moser, MB; Moser, EI; Forrest, E; Andersen, P; Morris, RG. Spatial learning with a minislab in the dorsal hippocampus. *Proc. Natl. Acad. Sci. USA,* 1995, 92, 9697-701.

Mundy, WR; Barone, S; Tilson, HA. Neurotoxic lesions of the nucleus basalis induced by colchicine: effects on spatial navigation in the water maze. *Brain Res.,* 1990, 512, 221-8.

Murray, CL; Fibiger, HC. Learning and memory deficits after lesions of the nucleus basalis magnocellularis: reversal by physostigmine. *Neuroscience,* 1985, 14, 1025-32.

Murray, EA; Davidson, M; Gaffan, D; Olton, DS; Suomi, S. Effects of fornix transection and cingulate cortical ablation on spatial memory in rhesus monkeys. *Exp. Brain Res.,* 1989, 74, 173-86.

Nagahara, AH; Otto, T; Gallagher, M. Entorhinal-perirhinal lesions impair performance of rats on two versions of place learning in the Morris water maze. *Behav. Neurosci.,* 1995, 109, 3-9.

Neave, N; Lloyd, S; Sahgal, A; Aggleton, JP. Lack of effect of lesions in the anterior cingulate cortex and retrosplenial cortex on certain tests of spatial memory in the rat. *Behav. Brain Res.,* 1994, 65, 89-201.

Nixon, PD; Passingham, RE. The cerebellum and cognition: cerebellar lesions do not impair spatial working memory or visual associative learning in monkeys. *Eur. J. Neurosci.,* 1999, 11, 4070-80.

Nonneman AJ; Kolb B. Functional recovery after serial ablation of prefrontal cortex in the rat. *Physiol. Behav.,* 1979, 22, 895-901.

Numan, R. Medial septal lesions impair performance on a preoperatively acquired delayed alternation task. *Brain Res. Bull.,* 1991, 26, 449-53.

Numan, R; Quaranta, JR Jr. Effects of medial septal lesions on operant delayed alternation in rats. *Brain Res.,* 1990, 531, 232-41.

Oades, RD; Isaacson, RL. The development of food search behavior by rats: the effects of hippocampal damage and haloperidol. *Behav. Biol.,* 1978, 24, 327-37.

O'Keefe, J; Dostrovsky, J. The hippocampus as a spatial map. Preliminary evidence from unit activity in the freely-moving rat. *Brain Res.,* 1971, 34, 171-5.

O'Keefe, J; Nadel, L; Keightley, S; Kill, D. Fornix lesions selectively abolish place learning in the rat. *Exp. Neurol.,* 1975, 48, 152-66.

O'Leary, TP; Brown, RE. Other mazes. In: WM Crusio, F Sluyter, RT Gerlai, S Pietropaolo (eds) *Behavioral genetics of the mouse,* vol 1. Cambridge: Cambridge University Press, pp 304-14, 2013.

Olton, DS; Papas BC. Spatial memory and hippocampal function. *Neuropsychologia,* 1979, 17, 669-82.

Oscar-Berman, M. The effects of dorsolateral-frontal and ventrolateral-orbitofrontal lesions on spatial discrimination learning and delayed response in two modalities. *Neuropsychologia,* 1975, 13, 237-46.

Packard, MG; Hirsh, R; White, NM. Differential effects of fornix and caudate nucleus lesions on two radial maze tasks: evidence for multiple memory systems. *J. Neurosci.,* 1989, 9, 1465-72.

Pandya, DN; Yeterian, EH. Proposed neural circuitry for spatial memory in the primate brain. *Neuropsychologia,* 1984, 222, 109-22.

Passingham, RE. Memory of monkeys (*Macaca mulatta*) with lesions in prefrontal cortex. *Behav. Neurosci.,* 1985, 99, 3-21.

Peternel, A; Hughey, D; Wenk, G; Olton, D. Basal forebrain and memory: neurotoxic lesions impair serial reversals of a spatial discrimination. *Psychobiology,* 1988, 16, 54-8.

Pioli, EY; Gaskill, BN; Gilmour, G; Tricklebank, MD; Dix, SL; Bannerman, D; Garner, JP. An automated maze task for assessing hippocampus-sensitive memory in mice. *Behav. Brain Res.,* 2014, 262, 249-57.

Pisa, M. Amnesia for discrete events in rats with fornicotomies: effects of interference on spatial alternation performance. *J. Comp. Physiol. Psychol.,* 1981, 95, 924-42.

Poucet B. Spatial cognitive maps in animals: new hypotheses on their structure and neural mechanisms. *Psychol. Rev.,* 1993, 100, 163-82.

Pribram, KH; Mishkin, M; Rosvold, HE; Kaplan, SJ. Effects of delayed-response performance of lesions of dorsolateral and ventromedial frontal cortex of baboons. *J. Comp. Physiol. Psychol.,* 1952, 45, 565-75.

Pritchett, K; Mulder, G. Hebb-Williams mazes. *Contemp. Top Lab. Anim. Sci.,* 2004, 43, 44-5.

Prusky, GT; West, PW; Douglas, RM. Reduced visual acuity impairs place but not cued learning in the Morris water task. *Behav. Brain Res.,* 2000, 116, 135-40.

Rabinovitch, MS; Rosvold, HE. A closed-field intelligence test for rats. *Can. Psychol.,* 1951, 5, 122-28.

Racine, RJ; Kimble, DP. Hippocampal lesions and delayed alternation in the rat. *Psychom. Sci.,* 1965, 3, 285-6.

Ramirez, JJ; Stein, DG. Sparing and recovery of spatial alternation performance after entorhinal cortex lesions in rats. *Behav. Brain Res.,* 1984, 1, 53-61.

Ramos, JMJ. Retrograde amnesia for spatial information: a dissociation between intra and extramaze cues following hippocampus lesions in rats. *Eur. J. Neurosci.,* 1998, 10, 3295-3301.

Rawlins, JNP; Olton, DS. The septo-hippocampal system and cognitive mapping. *Behav. Brain Res.,* 1982, 5, 331-58.

Richmond, MA; Yee, BK; Pouzet, B; Veenmam, L; Rawlins, JN; Feldon, J; Bannerman, DM. Dissociating context and space within the hippocampus: effects of complete, dorsal, and ventral excitotoxic hippocampal lesions on conditioned freezing and spatial learning. *Behav. Neurosci.,* 1999, 113, 1189-203.

Riekkinen, P Jr; Sirviö, J; Riekkinen, P. Similar memory impairments found in medial septal-vertical diagonal band of Broca and nucleus basalis lesioned rats: are memory defects induced by nucleus basalis lesions related to the degree of non-specific subcortical cell loss? *Behav. Brain Res.,* 1990a, 37, 81-8.

Riekkinen, P Jr; Sirviö, J; Riekkinen, P. Interaction between raphe dorsalis and nucleus basalis magnocellularis in spatial learning. *Brain Res.,* 1990b, 527, 342-5.

Roberts, WA. Spatial memory in the rat on a hierarchical maze. *Learn. Motiv.,* 1979, 10, 117-40.

Roberts, WW; Dember, WN; Brodwick, M. Alternation and exploration in rats with hippocampal lesions. *J. Comp. Physiol. Psychol.,* 1962, 55, 695-700.

Rochefort, C; Arabo, A; André, M; Poucet, B; Save E; Rondi-Reig, L. Cerebellum shapes hippocampal spatial code. *Science,* 2011, 334, 385-9.

Rodriguez, M; Gomez, C; Alonnso, J; Afonso D. Laterality, alternation, and perseveration relationships on the T-maze test. *Behav. Neurosci.,* 1992, 106, 974-80.

Rogers, JL; Kesner, RP. Lesions of the dorsal hippocampus or parietal cortex differentially affect spatial information processing. *Behav. Neurosci.,* 2006, 120, 852-60.

Rosenstock, J; Field, TD; Greene, E. The role of mamillary bodies in spatial memory. *Exp. Neurol.,* 1977, 55, 340-52.

Rosvold, HE; Mishkin, M; Szwarcbart, MK. Effects of subcortical lesions in monkeys on visual-discrimination and single-alternation performance. *J. Comp. Physiol. Psychol.,* 1958, 51, 437-44.

Rosvold, HE; Szwarcbart, MK; Mirsky, AF; Mishkin, M. The effect of frontal-lobe damage on delayed-response performance in chimpanzees. *J. Comp. Physiol. Psychol.,* 1961, 54, 368-74.

Salamone, JD; Beart, PM; Alpert, JE; Iversen, SD. Impairment in T-maze reinforced alternation performance following nucleus basalis magnocellularis lesions in rats. *Behav. Brain Res.,* 1984, 13, 63-70.

Saravis, S; Sziklas, V; Petrides, M. Memory for places and the region of the mamillary bodies in rats. *Eur. J. Neurosci.,* 1990, 2, 556-64.

Save, E; Moghaddam, M. Effects of lesions of the associative parietal cortex on the acquisition and use of spatial memory in egocentric and allocentric navigation tasks in the rat. *Behav. Neurosci.,* 1996, 110, 74-85.

Save, E; Poucet, B. Involvement of the hippocampus and associative parietal cortex in the use of proximal and distal landmarks for navigation. *Behav. Brain Res.,* 2000, 109, 195-206.

Schenk, F; Morris, RGM. Dissociation between components of spatial memory in rats after recovery from the effects of retrohippocampal lesions. *Exp. Brain Res.,* 1985, 58, 11-28.

Schneider, GE. Two visual systems. *Science,* 1969, 163, 895-902.

Schnerla, TC. Modifiability in insect behavior. In: KD Roeder (ed) *Insect physiology*. New York: John Wiley & Sons, pp 723-47, 1953.

Schwartzbaum, JS; Donovick, PJ. Discrimination reversal and spatial alternation associated with septal and caudate dysfunction in rats. *J. Comp. Physiol. Psychol.,* 1968, 65, 83-92.

Segal, M; Greenberger, V; Pearl, E. Septal transplants ameliorate spatial deficits and restore cholinergic functions in rats with a damaged septo-hippocampal connection. *Brain Res.,* 1989, 500, 139-48.

Shoji, H; Hagihara, H; Takao, K; Hattori, S; Miyakawa, T. T-maze forced alternation and left-right discrimination tasks for assessing working and reference memory in mice. *J. Vis. Exp.,* 2012, Feb 26:60.

Skelton, RW; McNamara, RK. Bilateral knife cuts to the perforant path disrupt spatial learning in the Morris water maze. *Hippocampus,* 1992, 2, 73-80.

Steffenach, H-A; Sloviter, RS; Moser, EI; Moser, M-B. Impaired retention of spatial memory after transection of longitudinally oriented axons of hippocampal CA3 pyramidal cells. *Proc. Natl. Acad. Sci. USA,* 2002, 99, 3194-8.

Stevens, R; Cowey, A. Effects of dorsal and ventral hippocampal lesions on spontaneous alternation, learned alternation and probability learning in rats. *Brain Res.,* 1973, 52, 203-24.

Stokes, KA; Best, PJ. Mediodorsal thalamus lesions impair radial maze performance in rats. *Behav. Neurosci.* 1988, 102, 294-300.

Stokes, KA; Best, PJ. Mediodorsal thalamus lesions in rats impair radial-arm maze performance in a cued environment. *Psychobiology,* 1990, 18, 63-7.

Stubley-Weatherly, L; Harding, JW; Wright JW. Effects of discrete kainic acid-induced hippocampal lesions on spatial and contextual learning and memory in rats. *Brain Res.,* 1996, 716, 29-38.

Sutherland, RJ; McDonald, RJ. Hippocampus, amygdala, and memory deficits in rats. *Behav. Brain Res.,* 1990, 37, 57-79.

Sutherland, RJ; Rodriguez, AJ. The role of the fornix/fimbria and some related subcortical structures in place learning and memory. *Behav. Brain Res.,* 1989, 32, 265-77.

Sutherland, RJ; Kolb, B; Whishaw, IQ. Spatial mapping: definitive disruption by hippocampal or medial frontal cortical damage in the rat. *Neurosci. Lett.,* 1982, 31, 271-76.

Sutherland, RJ; Kolb, B; Whishaw, IQ. A behavioural analysis of spatial localization following electrolytic, kainate- or colchicine-induced damage to the hippocampal formation in the rat. *Behav. Brain Res.,* 1983, 7, 133-53.

Sutherland, RJ; Kolb, B; Whishaw, IQ. Contributions of cingulate cortex to two forms of spatial learning and memory. *J. Neurosci.,* 1988, 8, 1863-72.

Sziklas, V; Petrides, M. Memory impairments following lesions to the mammillary region of the rat. *Eur. J. Neurosci.,* 1993, 5, 525-40.

Tako, AN; Béracochéa, DJ; Jaffard, R. Accelerated rate of forgetting spatial information following mamillary body lesions in mice: effects of context change on retention-test performance. *Psychobiology,* 1988, 16, 45-53.

Thinus-Blanc, C; Save, E; Rossi-Arnaud, C; Tozzi, A; Ammassari-Teule, M. The differences shown by C57BL/6 and DBA/2 inbred mice in detecting spatial novelty are subserved by a different hippocampal and parietal cortex interplay. *Behav. Brain Res.,* 1996, 80, 33-40.

Thomas, GJ; Spafford, PS. Deficits for representational memory induced by septal and cortical lesions (singly and combined) in rats. *Behav. Neurosci.,* 1984, 98, 394-404.

Thomas, GJ; Brito, GNO. Recovery of delayed alternation in rats after lesions of medial frontal cortex and septum. *J. Comp. Physiol. Psychol.,* 1980, 94, 808-18.

Thomas, GJ; Brito, GNO; Stein, DP. Medial septal nucleus and delayed alternation in rats. *Physiol. Psychol.,* 1980, 8, 467-72.

Thomas, JB. Stimulus perseveration and choice behavior in rats with septal lesions. *J. Comp. Physiol. Psychol.,* 1972, 80, 97-105.

Thomas, RK; Weir, VK. The effects of lesions in the frontal or posterior association cortex of rats on maze III. *Physiol. Psychol.,* 1975, 3, 210-4.

Thompson, R. Localization of the "maze memory system" in the white rat. *Physiol. Psychol.,* 1974, 2, 1-17.

Thompson, R. Rapid forgetting of a spatial habit in rats with hippocampal lesions. *Science,* 1981, 212, 959-60.

Thompson, R. Abnormal learning and forgetting of individual spatial reversal problems in brain-damaged rats. *Physiol. Psychol.,* 1983, 11, 35-46.

Thompson, R; Huestis, PW; Crinella, FM; Yu, J. The neuroanatomy of mental retardation in the white rat. *Neurosci. Biobehav. Rev.,* 1986, 10, 317-38.

Thompson, R; Bjelajac, VM; Huestis, PW; Crinella, FM; Yu, J. Inhibitory deficits in rats rendered "mentally retarded" by early brain damage. *Psychobiology,* 1989, 17, 61-76.

Thornton, EW; Davies, C. A water-maze discrimination learning deficit in the rat following lesion of the habenula. *Physiol. Behav.,* 1991, 49, 819-22.

Traverse, J; Latto, R. Impairments in route negotiation through a maze after dorsolateral frontal, inferior parietal or premotor lesions in cynomolgus monkeys. *Behav. Brain Res.,* 1986, 20, 203-15.

Turnbull, J; Jiang, F; Racine, R. Hippocampal stimulation of fornical-lesioned rats improves working memory. *Can. J. Neurol. Sci.,* 1994, 21, 100-3.

Ulrich, K; Aitken, PN; Abraham, WC; Dalrymple-Alford, JC; McNaughton, N. Effects of thalamic lesions on repeated relearning of a spatial working memory task. *Behav. Brain Res.,* 2014, 261, 56-9.

Van der Staay, FJ; Raaijmakers, WG; Lammers, AJ; Tonnaer, JA. Selective fimbria lesions impair acquisition of working and reference memory of rats in a complex spatial discrimination task. *Behav. Brain Res.,* 1989, 32, 151-61.

Van Haaren, F; de Bruin, JPC; Heinsbroek, RP; Van de Poll, NE. Delayed spatial response alternation: effects of delay-interval duration and lesions of the medial prefrontal cortex on response accuracy of male and female Wistar rats. *Behav. Brain Res.,* 1985, 18, 41-9.

Vann, SD; Kristina Wilton, LA; Muir, JL; Aggleton, JP. Testing the importance of the caudal retrosplenial cortex for spatial memory in rats. *Behav. Brain Res.,* 2003, 140, 107-18.

von Frisch, K. *The dance language and orientation of bees.* Cambridge, MA: Belknap Press of Harvard University Press, 1967.

Vorhees, CV. Maze learning in rats: a comparison of performance in two water mazes in progeny prenatally exposed to different doses of phenytoin. *Neurotoxicol. Teratol.,* 1987, 9, 235-41.

Walker, JA. Common components of information processing underlying memory disorders in humans and animals. In: HL Roitblat, TG Bever, HS Terrace (eds) *Animal cognition.* Hillsdale: Lawrence Erlbaum, pp 591-605, 1984.

Walker, JA; Olton, DS. Spatial memory deficit following fimbria-fornix lesions: independent of time for stimulus processing. *Physiol. Behav.,* 1979, 23, 11-5.

Walker, JA; Olton, DS. Fimbria-fornix lesions impair spatial working memory but not cognitive mapping. *Behav. Neurosci.,* 1984, 98, 226-42.

Wan, R-Q; Pang, K; Olton, DS. Hippocampal and amygdaloid involvement in nonspatial and spatial working memory in rats: effects of delay and interference. *Behav. Neurosci.,* 1994, 108, 866-82.

Warburton, EC; Aggleton, JP. Differential deficits in the Morris water maze following cytotoxic lesions of the anterior thalamus and fornix transection. *Behav. Brain Res.,* 1999, 98, 27-38.

Warburton, EC; Aggleton, JP; Muir, JL. Comparing the effects of selective cingulate cortex lesions and cingulum bundle lesions on water maze performance by rats. *Eur. J. Neurosci.,* 1998, 10, 622-34.

Weis, BJ; Means, LW. A comparison of the effects of medial frontal, dorsomedial thalamic and combination lesions on discrimination and spontaneous alternation in the rats. *Physiol. Psychol.,* 1980, 8, 325-9.

Whishaw, IQ. Formation of a place learning-set by the rat: a new paradigm for neurobehavioral studies. *Physiol. Behav.,* 1985, 35, 139-43.

Whishaw, IQ; Maaswinkel, H. Rats with fimbria-fornix lesions are impaired in path integration: a role for the hippocampus in "sense of direction". *J. Neurosci.,* 1998, 18, 3050-8.

Whishaw, IQ; Maaswinkel, H; Gonzalez, CL; Kolb, B. Deficits in allothetic and idiothetic spatial behavior in rats with posterior cingulate cortex lesions. *Behav. Brain Res.,* 2001, 118, 67-76.

Wikmark, RGE; Divac, I; Weiss, R. Retention of spatial delayed alternation in rats with lesions in the frontal lobes: implications for a comparative neuropsychology of the prefrontal system. *Brain Behav. Evol.,* 1973, 8, 329-39.

Wilton, LA; Baird, AL; Muir, JL; Aggleton, JP. Excitotoxic lesions of the rostral thalamic reticular nucleus do not affect the performance of spatial learning and memory tasks in the rat. *Behav. Brain Res.,* 2001a, 120, 177-87.

Wilton, LA; Baird, AL; Muir, JL; Honey, RC; Aggleton, JP. Loss of the thalamic nuclei for "head direction" impairs performance on spatial memory tasks in rats. *Behav. Neurosci.,* 2001b, 115, 861-9.

Wirtshafter, D; Asin, KE. Impaired radial arm maze performance in rats with electrolytic lesions of median raphe lesions. *Exp. Neurol.,* 1983, 79, 412-21.

Zola-Morgan, S; Squire, LR. Medial temporal lesions in monkeys impair memory on a variety of tasks sensitive to human amnesia. *Behav. Neurosci.,* 1985, 99, 22-34.

In: Spatial Orientation in Humans and Animals
Editor: Robert Lalonde
ISBN: 978-1-63463-001-6
© 2014 Nova Science Publishers, Inc.

Chapter 3

STRUCTURAL BASIS OF SPATIAL ORIENTATION

Catherine Strazielle*
Université de Lorraine, Laboratoire "Stress, Immunité, Pathogènes" and Service de Microscopie Electronique, Faculté de Médecine and CHU de Nancy, Vandoeuvre-les-Nancy, France

INTRODUCTION

Spatial orientation is a physiological process which permits us to interact with the environment or move in it. This cognitive function, linked to the sensori-motor system, depends on multiple cortical and subcortical networks which cooperate to produce a behavior pattern totally adapted to the environment. Therefore, spatial memory has the capacity to select the pertinent information about one's environment to establish multiple levels of its representation in the brain, under the form of cognitive maps, either egocentric (defining the self in relation to objects, goal, or intra-maze locations) or allocentric (defining the self in relation to the three-dimensional arrangement of objects in the environment) (Moscovitch et al., 2005).

This chapter outlines the anatomical substrate for spatial learning, planning, and acting in space. Multiple studies, referred in the other chapters, demonstrated that occipital, frontal, parietal, and temporal cortices with interconnecting regions such as the cerebellum and neostriatum are related to spatial cognition. We propose to study the neuroanatomical attainments of these various brain regions for a better comprehension of the functional networks and of the clinical symptoms and behavior alterations when those are impaired.

1. THE VISUAL SYSTEM

Spatial orientation depends on guide marks for which the visual system is essential by identifying, selecting, and recognizing the pertinent cues on a route while a person is moving. The visual system is organized into two major pathways (Ungerlieder and Mishkin, 1982),

* Email: catherine.strazielle@univ-lorraine.fr

their segregation occurring from retina to the occipital lobe, passing through the lateral geniculate nucleus of the thalamus: large retinal ganglion cells (M cells) project to the magnocellular layers of the lateral geniculate nucleus forming the dorsal M pathway, while the ventral P pathway originates with projections of small retinal ganglion cells (P cells) to the parvocellular layers of the thalamic nucleus, those two pathways projecting into multiple but distinct cortical areas (Clark et al., 2005; Ungerlieder and Mishkin, 1982).

The visual cortex corresponds to Brodmann areas (BA) 17, 18, and 19, located in the occipital lobe and overlapping on parietal and temporal lobes. More recently, Zeki (1993) described six visual areas (V1-V6), each with a specific role. The first projection is on the primary visual area (V1 = BA 17), the other areas are associative. The M stream forms the dorsal occipitoparietal pathway, from V1 to the posterior parietal cortex, including the middle temporal area (V5); it is involved in the recognition of spatial relations permitting an answer to the question of «where» a stimulus is and consequently in the visual guidance of behavior. The P stream forms the ventral occipitotemporal pathway, from V1 to V4 localized in the inferior temporal cortex; it is crucial for the identification of objects, permitting an answer to the question of «what» the stimulus is. Besides these direct pathways from the lateral geniculate nucleus, indirect pathways from the superior colliculus and pulvinar may reach the visual associative areas without a V1 projection (Clark et al., 2005). Although these pathways are separated, visual integration and perception are likely to be a multistage process requiring interconnections within and between the pathways. The major projections from the visual pathways reach the prefrontal cortex. Returned fronto-occipital fasciculi affect visual attention.

Particularly enlarged in the primate brain, the pulvinar is a thalamic region of the lateral nuclear group. As described above, it forms part of the secondary extrageniculostriate visual system. It is interconnected with the superior colliculus and projects principally into the posterior association cortices (Baleydier and Mauguiere, 1987). The superior colliculus is a subregion of the tectum, localized in the dorsal part of the midbrain. In inferior vertebrates, the superior colliculus is one of the most important centers to treat visual information. In mammals, following telencephalization, the occipital lobe supersedes it. The major central inputs of the superior colliculus arise from visual and multimodal associative areas and from the substantia nigra (Huerta and Harting, 1984). In association with the pulvinar and several cortical visuomotor regions localized in frontal and parietal lobes (Kaas, 1989), it is essential for the control of eye movements and visual attention.

2. Cortical Regions Involved in Spatial Orientation

The cerebral cortex is described as a mosaic of various functional areas which follow a hierarchical organization into primary sensory or motor areas and limbic areas, coupled with association areas (Clark et al., 2005). The latter ones are unimodal when they receive their main afferents from other cortical zones specialized in only one specific modality. They are linked with primary areas, lying besides them. In turn, they project into multimodal association areas which integrate information about multiple modalities to manage the highest brain functions such as spatial cognition. There are three multimodal association areas in the cerebral cortex: (1) the posterior somesthetic association area in the parieto-temporal region, localized at the margin of parietal, temporal, and occipital lobes, is concerned with visuospatial localization and attention, (2) the anterior motor association area, corresponding to prefrontal cortex, localized rostral to premotor areas, is concerned with planning movements and strategies, and (3) the limbic association area or paralimbic area, composed of cingulate gyrus, hippocampal formation, parahippocampal gyrus, and amygdala, is concerned with memory storage and emotions. Thus, the four cortical lobes are involved in spatial function. The occipital cortical regions as part of the visual system were already mentioned in the previous paragraph.

2.1. Parietal Lobe

The parietal lobe is bordered by the central sulcus ahead and lies on both sides of the cingulate gyrus. The parietal lobe consists of the primary somatosensory cortex in its anterior portion, corresponding to BA 1, 2, and 3 and the posterior parietal lobule divided into superior parietal lobule (BA 5 and 7) and inferior parietal lobule (BA 39 and 40), both separated, in primates, by the intraparietal sulcus (Clark et al., 2005).

The primary somatosensory cortex (SI) is the target of thalamic neurons relaying somesthetic information in ventral posteromedial and posterolateral nuclei. The somatotopic organization of BA 1-3 is multiple and variable according to the sensory modalities. In rodents, the barrels form a specific parietal zone arranged in rows similar to the topography of the whiskers (Palomero-Gallagher and Zilles, 2004). Cortical neurons project in return to the thalamic nuclei, to primary motor cortex, and mainly to the associative areas in the adjacent posterior parietal lobule. The unimodal associative areas consist of the anterior part of BA 5 and 7 in addition with the supplementary sensory area and secondary somatosensory area (SII, BA 43) lying at the foot of SI adjacent to the insula, the latter having the particularity to receive somesthetic information from both sides of the body. The multimodal associative areas are large, including the posterior part of BA 5 and 7 with the BA 39 and 40 (Clark et al., 2005). Highly developed in humans, the posterior parietal cortex contributes mainly to the formation of the multimodal posterior association area. In connection with several other cerebral regions, it is involved in higher cognitive functions.

The superior parietal lobule has reciprocal connections with the pulvinar and lateral posterior thalamic nucleus, and sends projections to motor control regions such as the red nucleus, basal ganglia, superior colliculus, pontine tegmentum, and cerebellum (Clark et al., 2005). The superior parietal lobule is the terminal projection region of the dorsal occipitoparietal pathway. Short association fibers link it with adjacent cortices concentrating visual, sensory, vestibular, and auditory inputs as well as signals from superior and middle temporal cortex (Clark et al., 2005; Lopez and Blanke, 2011) to create cognitive frames of the space surrounding the body. The processing of space and objects in space by visuospatial integration and attention (Colby and Goldberg, 1999) as well as preparation of orientated actions in space (Andersen and Cui, 2009; Bremmer, 2011; Sakata et al., 1997) are therefore the main functions of this lobule but, being lateralized, they divide unequally between both sides, the right cortex being the preferred one. Long association fibers interconnect the superior parietal lobule with the laterodorsal prefrontal, cingulate, and motor cortices such as the premotor, frontal eye field, and supplementary motor areas responsible for the spatial guidance of motor behavior in the body environment (Andersen and Cui, 2009; Fogassi and Luppino, 2005; Goodale and Milner, 1992). Such a multimodal convergence is also documented in the rat parietal cortex and its functional organization presents similarities since the ventrolateral part of secondary visual area (OC2L) is included in the posterior parietal lobule (Palomero-Gallagher and Zilles, 2004). Moreover, findings showed the existence of a specific visual processing pathway involved in visuospatial guidance in the rat posterior parietal cortex (Kolb and Walkey, 1987; Reep et al., 1994).

Described in primates, the region of the intraparietal sulcus that runs between the superior and inferior parietal lobule also contributes to spatial cognitive functions. It is divided in functionally distinct intraparietal areas: neurons in the ventral intraparietal area may encode visual and somatosensory information (Avillac et al., 2007; Guipponi et al., 2013) in a head-centered representation of space for guiding head movements; neurons in the lateral intraparietal area, receiving projections from visual areas, may be involved in spatial attention (Bisley and Goldberg, 2003), selecting the relevant visual stimuli and controlling eye movements with other brain centers such as the frontal eye field, basal ganglia, and superior colliculus (Yantis et al., 2002); the medial intraparietal area may be a source of spatial information used by the premotor cortex for reaching specific targets (Caminiti et al., 1996).

The inferior parietal lobule also has reciprocal connections with the superior colliculus, the pulvinar and other lateral thalamic nuclei. Short association fibers interconnect it with the superior parietal lobule and the contiguous occipital and temporal lobes. Long association fibers interconnect it with the multimodal sensory convergence area in the temporal lobe, the ventromedial temporal lobe, the frontal lobe, including the frontal eye field and the dorsolateral prefrontal cortex as well as the cerebellum (Clark et al., 2005; Clower et al., 2001).

2.2. Temporal Lobe

The temporal lobe is a complex cortical structure divided into a lateral neocortical region and a ventromedial limbic region. It lies rostral to the occipital lobe and ventral to the parietal lobe from which it is separated by the lateral fissure. The lateral region, partly involved in spatial cognition, is the focus of this paragraph. It is superficially organized in auditory cortex and temporal associative areas. The auditory cortex occupies the superior surface of the superior temporal gyrus corresponding to the transverse gyri. BA 41 and 42 are respectively the primary and secondary auditory cortices receiving their main afferents from the medial geniculate nucleus.

They are surrounded by the auditory association area named 22 by Brodmann and localized in the superior temporal gyrus. All the auditory cortices establish reciprocal connections with frontal, parietal or temporal association areas (Clark et al., 2005). The temporal associative areas occupy the middle temporal gyrus (BA 21 and 37 caudally) and the inferior temporal gyrus (BA 20 and 37 caudally), involved with visual processing, all the more since the inferior temporal gyrus extends to the inferior surface of the lobe to form caudally the lateral occipitotemporal or fusiform gyrus, site of visual areas V4 and V5 which correspond with the terminal projection region of the ventral occipitotemporal pathway (Ungerleider and Mishkin, 1982).

The cortex deep within the superior temporal sulcus concentrates multiple sensory modalities, vision, somatic sensation, and audition with limbic inputs. This convergence area is in continuity with the posterior parietal cortex to constitute the multimodal posterior association area already described. Efferents project to associative areas of frontal and parietal cortices as well as limbic-related structures, thalamic associative nuclei, particularly the mediodorsal and pulvinar, and the basal ganglia (Clark et al., 2005).

2.3. Frontal Lobe

The frontal cortex is involved in spatial functions as a logical result of its fundamental role in the control of motor behavior at all levels. Anatomically, the frontal lobe is composed of three distinct regions, dorsolateral, medial, and ventral. Functionally, it can be divided into (1) a motor cortex in the posterior part of both the dorsolateral and medial regions, composed of a primary motor and associative areas, and (2) a prefrontal cortex in rostral position, involved in a higher level of spatially related functions including memory.

Among the motor cortex, the primary motor cortex (M1 = BA 4) corresponding to the precentral gyrus gives rise to the major lateral descending motor system, projecting directly to the ventral horn of the spinal cord (pyramidal tract) and cranial nerve motor nuclei of the brainstem (geniculate tract) or indirectly via the red nucleus. Fibers from premotor areas and parietal somesthetic cortex also contribute to the efferent tracts (Clark et al., 2005). The primary motor cortex is involved in space insofar as it contains neurons selective for movement direction (Kakei et al., 1999), whose control is provided by cortical loops of connections with the striatum and the cerebellum. Among the associative motor cortex localized in front of M1, the BA 6 is divided into premotor area and supplementary motor area, respectively on lateral and superomedial sides of the lobe. The premotor cortex receives definite topographically organized inputs from the superior parietal cortex (Wise et al., 1997). It also receives projections from other anterior prefrontal and posterior associative zones as well as from primary motor cortex (Clark et al., 2005; Stuss and Benson, 1986). Efferent fibers that leave the premotor cortex terminate, on one hand, in the primary motor cortex, and on the other hand, in striatum and cerebellum, red nucleus, thalamus, and reticular formation of the pons and medulla which gives rise to the reticulospinal tracts, part of the major medial descending motor system in support of body posture and locomotion (Clark et al., 2005; Stuss and Benson, 1986). Premotor areas are activated for the initiation or modification of motor programs and particularly when movements are visually guided. Supplementary motor area connections are similar but with additional afferents arising from the prefrontal cortex, particularly the anterior cingulate gyrus, regions producing cognitive and motivational information (Picard and Strick, 1996). Some authors subdivide this supplementary motor area into two parts, the rostral pre-supplementary one projecting to the caudal supplementary, more related to movement execution and in relation with the premotor area (Picard and Strick, 1996; Rizzolatti et al., 1996). Studies indicate that the rostral pre-supplementary motor area becomes active in preparation or even thinking of cued movements, thus indicating its involvement in motor planning and possibly in procedural memory (Grafton et al., 1992; Sahyoun et al., 2003). The frontal eye field on the dorsolateral frontal cortex occupies the major part of BA 8. Interconnected with the posterior parietal cortex, more precisely the lateral intraparietal area, as well as with the visual areas and superior colliculus, it also projects to caudate nucleus, cerebellum, and pontine reticular formation (Clark et al., 2005; Sommer and Wurth, 1998; Stuss and Benson, 1986). The frontal eye field contributes to attention and oculomotor functions, ordering the brainstem to move the eyes on one target selected among many (Buttner-Ennever, 1988).

The prefrontal cortex is specially developed in primates, even more so in humans. It models itself on the entire lobe, consisting of three topographical zones, dorsolateral, medial, and ventral (ventrolateral and orbital areas) functionally distinct throughout their connections in action planning and working memory, motivation, and emotion, respectively (Roberts et al., 2000). They all receive projections from the thalamic mediodorsal nucleus associated with the limbic system (Ray and Price, 1993). Our focus is the dorsolateral part of the frontal cortex, in charge of spatially related functions, known as BA 9 and 46. The frontal eye field (BA 8) is considered as an association region because of an overlapping of its anterior portion with those prefrontal areas. This cortical region is extensively connected bidirectionally with motor cortex, cingulate cortex, and cortical sensory association areas (Clark et al., 2005; Roberts et al., 2000). It receives sensory information from multiple modalities (visual, auditory, and somato-sensory) that constantly updates it on the state of the environment. By receiving large visuospatial information from the posterior parietal area (BA 7 and lateral part of intraparietal areas), it may be considered as the efferent pole of the dorsal visual stream, producing spatial attributes of working memory and motor planning (Cavada and Goldman-Rakic, 1989; Katsuki and Constandinidis, 2012; Selemon and Goldman-Rakic, 1988). The laterodorsal prefrontal cortex shares many efferent targets with the posterior parietal cortex such as premotor, supplementary motor, superior temporal, and cingulate cortices, limbic structures, thalamus, and superior colliculus (Katsuki and Constandinidis, 2012; Selemon and Goldman-Rakic, 1988). They are major sources of projections to the neocerebellum via pontine efferents (Schmahmann and Pandya, 1997) which in turn sends efferent fibers back to the cortex by way of the thalamus. They are implicated in similar loops with the dorsal neostriatum, cortical afferents ending particularly on cells in the dorsomedial striatum (caudate nucleus) involved in motor learning and strategies. Dopaminergic innervation is predominant in the frontal lobe (Haber and Fudge, 1997; Levitt et al., 1984). It is recognized as a critical factor of prefrontal functions, the D1 receptors being widely spread in the region, particularly in its medial part (Lewis, 1988). The dopaminergic innervation may play a role in the motor cognitive function because the injection of D1 antagonists in the laterodorsal prefrontal impairs the performances in working memory and attention tasks (Montague et al., 2004; Noudoost and Moore, 2011; Sawaguchi and Goldman-Rakic, 1991).

3. The Hippocampus and Related Limbic Structures

The hippocampus is generally regarded as the main structure underlying learning and memory, including spatial material. By encoding complex spatial information, the hippocampus plays a crucial role in the formation and retention of cognitive allocentric maps of the environment (Jacobs and Schenk, 2003; Squire, 1992). Therefore, navigation capacity, tested in numerous tasks, serves to evaluate hippocampal integrity in animals. It is an early function in vertebrate evolution because the telencephalic pallial areas of reptiles and fish, the presumed homologues of the mammalian hippocampus, are similarly involved in spatial navigation (Rodriguez et al., 2002).

The hippocampus, part of the limbic system, is localized in the ventromedial portion of the temporal lobe. Its terminology is variable and must be explained. The hippocampus proper includes Ammon's horn (CA = *cornu ammonis*) segmented into three pyramidal cell fields (CA1-CA3). Sometimes, four pyramidal cell fields (CA1-CA4) are described but the term CA4 was improperly attributed to cells of the polymorphic cell layer of the dentate gyrus because of a confusion with the terminal portion of the hippocampal pyramidal cell layer (Amaral, 1978; Amaral and Witter, 1995). We have chosen to use the term hippocampal formation for the allocortex, consisting only of the dentate gyrus, principal region receiving afferents, the hippocampus proper, and the subiculum, principal source of efferent fibers out of the hippocampus. The region is called hippocampal complex when the hippocampal formation includes the adjacent parahippocampal region which is composed of entorhinal (BA 28), perirhinal (BA 36), and postrhinal (BA 35) cortices, along with the presubiculum (BA 27) and parasubiculum (BA 49) (for a detailed systematization according to animal species, see Witter et al., 1989; Zola-Morgan and Squirre, 1993). Sometimes, the parasubiculum and presubiculum are improperly grouped together with the subiculum as a subicular complex when they are input structures with a cytoarchitectonic organization similar to the entorhinal cortex rather than output structures like the subiculum (Amaral and Witter, 1995; Lopes da Silva et al., 1990). The hippocampal formation in association with the cingulate and parahippocampal gyri were described by Broca as the limbic lobe («le grand lobe limbique»), interconnected cortices acting together as a functional entity.

3.1. The Hippocampal Complex

The hippocampal formation, principal structure of the limbic association area, receives a major cortical input, particularly from the limbic cortex and the multimodal association areas of the parieto-temporal and frontal lobes. However, since it is an old cortex (allocortex) with three fundamental layers, these input are indirect, passing through the interface of the parahippocampal region, whose layered-cellular organization forms a transitional cortex. Thus, the entorhinal cortex, central part of the parahippocampal region, is considered as the gateway to the hippocampal formation (Amaral and Witter, 1995; Insausti et al., 1997). Entorhinal fibers originating from the superficial layer II make up the perforant path that terminates preferentially in the dentate gyrus, but also in the CA fields and subiculum (Figure 1). More broadly presented, the overall areas of the parahippocampal region constitute the major afferent of the hippocampal formation, indirectly by projections first into the entorhinal cortex, or directly, in a minor proportion, into the CA1 subfield and subiculum of the hippocampal formation (Amaral and Witter, 1995). Other subcortical sources of input, distributing similarly to all hippocampal sectors, include the septal region, the amygdala, the posterior hypothalamus, the anterior, paraventricular, and midline nuclei of the thalamus, as well as monoaminergic nuclei of the brainstem, namely the raphe nuclei, locus coeruleus, and ventral tegmental area (Amaral and Witter, 1995). Inside the hippocampal formation, a synaptic chain links one after another the dentate gyrus, CA3, CA1, and subiculum. This well-ordered pattern of connectivity is of functional importance, determining the flow by which information is processed. Moreover, besides afferent projections of the perforant fibers, reciprocal connections, originating in CA1 subfield and subiculum, project directly or indirectly by way of the presubiculum, parasubiculum, or perirhinal area, in the deep cell layers of the entorhinal cortex (Witter et al., 1989). Thus, several parallel but strongly interconnected circuits exist between the parahippocampal region, more specifically the entorhinal cortex and the hippocampal formation (Naber et al., 2001). Functioning as reverberatory circuits, these feedback pathways may facilitate memory function by lowering the threshold of entorhinal neurons to recognize more rapidly a pattern of sensory signals (Buzsaki et al., 1990; Deadwyler et al., 1975; Van Groen and Lopes da Silva, 1986).

In addition to these strong hippocampal-parahippocampal efferents, the output from the subiculum projects directly to the medial frontal cortex and the caudal cingulate cortex (Van Hoesen, 1982; Witter et al., 1989). The other outgoing projections from the hippocampal formation arise from axons of the subiculum and hippocampal pyramidal cells. They form first the fimbria, then the fornix which projects to various subcortical regions, mainly the septal nuclei, hypothalamus and mammillary bodies, and in weaker proportion the midline and anterior thalamus, the striatum, the amygdala, and the midbrain tegmentum (Amaral and Witter, 1995).

A specific loop composed of the hippocampus, fornix, mammillary bodies, mammillothalamic tract, anterior thalamic nuclei, cingulate gyrus, and projections back to the hippocampus by way of the entorhinal cortex (Figure 1) form the Papez circuit (1937). Considered at first as a central emotional circuitry, it is now recognized that the hippocampal-cingulate network passing through the anterior thalamus is involved in mnemonic and attentional processes (Aggleton and Brown, 1999; Mega et al., 1997; Shah et al., 2012; Vertes et al., 2001). The recent description of several parallel hippocampal-anterior thalamic circuits with different roles underscores the pivotal role of the anterior thalamus in the Papez circuit and spatial navigation (Jankowski et al., 2013). The feedback pathways from the cingulate gyrus and anterior thalamus to the hippocampal complex could be considered as a mnemonic circuit of reinforcement.

We have seen that the hippocampal formation and the adjacent parahippocampal region form strong interconnections. Studies in cat and monkey showed that the entorhinal cortex and other parts of the parahippocampal region are reciprocally connected with widespread areas of the telencephalic cortex, including most of the limbic cortex and extensive parts of the association cortex (Lopes da Silva et al., 1990; Lavenex et Amaral, 2000; Van Hoesen, 1982; Witter et al., 2000). Thus, the parahippocampal region provides the crucial link between the hippocampal formation and the neocortex with which it is not directly connected (Parron et al., 2006). These cortical sources represent the major substrate by which the hippocampal complex builds spatial maps constituted by a bound ensemble of neurons specifically activated for one's environment. Pyramidal cells of the hippocampus, such as «place cells» as well as specific « grid cells » in the entorhinal cortex, fire more actively when an animal is in a specific area referred to as a place field to encode the salient objects in its sourroundings and build global topographic cartographies of them (O'Keefe and Dostrovsky 1971; Silva et al., 1998). In parallel, electrophysiological studies showed that these hippocampal neuronal representations of space are strongly influenced by vestibular information (Good, 2002). The different subfields of the hippocampus may subserve different functions, the entorhinal-CA1 circuit being important for learning, whereas an intact CA3-CA1 connectivity is necessary for recall (Brun et al., 2002). Studies in rodents or humans suggest that activations associated with memory encoding are located primarily in the anterior portion of the hippocampus, whereas retrieval activations are located more posteriorly (Lepage et al., 1998; Maguire et al., 1999; Moser and Moser, 1998).

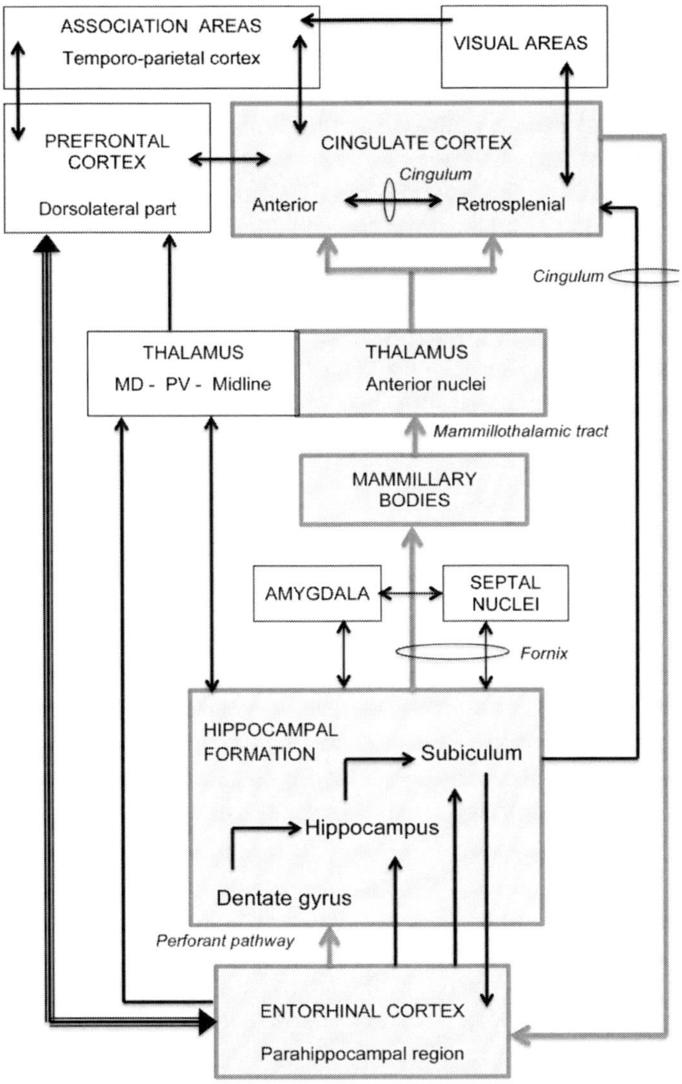

Figure 1. Schematic diagram of the extended connections of the hippocampus complex. By way of the parahippocampal region, mainly composed of the entorhinal cortex, the hippocampal formation has strong interconnections with the multimodal association areas of parieto-temporal and frontal lobes. Entorhinal efferents to the hippocampal formation form the perforant path. Major hippocampal efferents project to amygdala, hypothalamus (mammillary bodies), and septum. The Papez circuit is highlighted in gray: the fornix connects the hippocampus with the mammillary bodies; the mammillothalamic tract ascends to the anterior thalamic nuclei; the cingulum contains the afferents fibers to the cingulate cortex and the feedback efferents fibers from the cingulate to the entorhinal cortex. Besides the Papez circuit, the hippocampal complex sends efferents to the retrosplenial cortex intimately connected with visual areas and to the prefrontal cortex via the mediodorsal nucleus of the thalamus. In the cerebral cortex, the prefrontal cortex is extensively linked with the somesthetic association areas; the cingulate cortex and the prefrontal cortex are interrelated, many authors including the anterior cingulate area as part of the dorsomedial prefrontal area.

3.2. The Cingulate Cortex

The cortical area lying over the parahippocampal gyrus is divided into anterior cingulate (BA 24) and posterior cingulate (BA 23), which includes the retrosplenial cortex (BA 29) in its more caudal portion. Many neuroanatomical observations underscore the significant involvement of the retrosplenial cortex in visuospatial orientation and memory (Aguire and D'Esposito, 1999; Sutherland and Hoesing, 1993): (1) it receives visual input from the pulvinar, visual associative areas, and prefrontal cortex (Kobayashi and Amaral, 2003), (2) it has strong interconnections with the hippocampal complex, specially the entorhinal cortex, subiculum, and presubiculum, the latter rich in visual input (Amaral and Witter, 1995; Kobayashi and Amaral, 2003), (3) it has strong interconnections with the anterior thalamus, part of the Papez circuit, all these connections providing the relevant information code for head and body position in allocentric space (Shibata, 1993; Taube et al., 1990). The retrosplenial cortex is involved in the performance of learned tasks by promoting orienting reponses to novel target and recall of the past location of significant targets (Sutherland and Hoesing, 1993). Visuospatial impairment was coupled with a significant reduction of cerebral blood flow in the posterior cingulate cortex (Elfgren et al., 2003). Such symptoms can occur two years previous to the development of Alzheimer's disease (Huang et al., 2002).

Some efferent fibers from the posterior cingulate cortex project directly to the striatum and to the pontine nuclei but the major outputs take the cingulum to terminate in the cingulate motor area localized in the dorsal part of the anterior cingulate cortex (Figure 1). This motor area is connected with numerous motor structures such as the supplementary motor area, frontal eye field, primary motor cortex, striatum, and spinal cord (Bates and Goldman-Rakic, 1993; Clark et al., 2005), attesting the importance of the cingulate cortex in motor behavior and more precisely in attention and orientation to external stimuli.

3.3. The Cholinergic System

The cholinergic system has preferential relationships with spatial cognition (Deiana et al., 2011), both features known to be impaired in Alzheimer's disease (Schliebs and Arendt, 2011). The basal forebrain contains large cholinergic neurons scattered in the magnocellular basomedial telencephalic complex, extending in the rostral to caudal direction from the olfactory tubercle to the hypothalamic region.

Cholinergic neurons are grouped into 6 subareas (CH1-CH6) according to the nomenclature of Mesulam and colleagues (1983), the first four groups localized in the basal forebrain. The medial septum (Ch1) as well as vertical and horizontal limbs (Ch2 and Ch3, respectively) of the diagonal band of Broca forms the septal nuclear complex. The last forebrain cholinergic area (Ch4), located in substantia inominata/ventral pallidum in rodents, equivalent to the nucleus basalis magnocellularis (= basal nucleus of Meynert) in primates, overlaps in the medullary laminae of the pallidum. There are also two midbrain cholinergic regions (CH5 and CH6) consisting of pedunculopontine and laterodorsal tegmentum, respectively, and their projections mainly target cortical and various thalamic regions (Everitt and Robbins, 1997).

Cholinergic innervation exerts its role in spatial cognition at different functional levels of the brain. The hippocampus receives its projecting fibers from septal nuclei CH1 and CH2 by way of the septo-hippocampal pathway; the level of acetylcholine may be high during spatial acquisition and low during consolidation (for references, see Deiana et al., 2011); a cholinergic septo-entorhinal projection may also be involved in learning and memory (Walsh et al., 1996). The prefrontal cortex may solicit the activity of its projecting cholinergic fibers from the nucleus basalis during working memory (for references, see Deiana et al., 2011). Although dopamine plays a crucial role in the functioning of the basal ganglia, the striatal interneurons are cholinergic. The pedonculopontine tegmental nucleus and the dorsal tegmental nucleus are known to play a role in attention, arousal, and spatial learning by way of their projections to specific nuclei of the basal ganglia and thalamus as well as the mammillary bodies (Dwyer et al., 2013; Steckler et al., 1994; Winn et al., 1997).

4. The Basal Ganglia

Traditionally regarded as a motor control system, the basal ganglia provide one of the loop systems of the neocortex involved in generating motor behavior but also cognitive and emotional functions. The basal ganglia include (1) the corpus striatum divided into the neostriatum (putamen, caudate nucleus, as dorsal striatum and nucleus accumbens as ventral striatum) and the paleostriatum composed of medial (GPi) and lateral (GPe) segments of the globus pallidus followed ventrally by the ventral pallidum or substantia inominata, (2) regions closely tied to the previous structures namely the subthalamic nucleus, and the midbrain dopaminergic region made up of the substantia nigra and the ventral tegmental area (Lalonde and Strazielle, 2010).

The basal ganglia represent the intrinsic nuclei of distinct motor, oculomotor, associative, and limbic circuits organized in cortico-striato-cortical loops running parallel to each other (Alexander and Crutcher, 1990). While the three former circuits are associated with the dorsal striatopallidal system, the latter one, producing motivational aspects of motor behavior, is associated with the ventral striatopallidal system. The cortical inputs are abundant glutamatergic fibers from all portions of the cerebral cortex projecting in accordance with a functional somatotopy: fibers from the frontal motor areas and primary somatosensory cortex end on putamen cells involved in motor loops; fibers from the dorsolateral prefrontal cortex and somesthetic association areas terminate on caudate cells involved in the associative loop; fibers from the hippocampal complex and amygdala end on cells of the nucleus accumbens involved in the limbic loop (Clark et al., 2005). The dorsal striatum and more specifically the caudate nucleus is an important region in cognitive function (Graybiel, 1997) and involved in spatial learning (Iaria et al., 2003; McDonald and White, 1994). Additional afferent fibers to the basal ganglia arise from the intralaminar thalamic nuclei, the dopaminergic nucleus of the midbrain, and the raphe nuclei.

Two main pathways, one direct and the other indirect (passing through the external globus pallidus and the subthalamic nucleus) connect the striatum with the output nuclei of the basal ganglia, namely the medial globus pallidus and the substantia nigra pars reticulata.

These outputs are directed to the ventral anterior and ventral lateral thalamic nuclei with collaterals to the centromedian nucleus except for the limbic loop, whose output fibers can relay in the mediodorsal thalamic nucleus. The thalamic nuclei provide feedback projections to the cerebral cortex: it is generally admitted that the main cortical targets of the basal ganglia outputs are limited to the supplementary motor and lateral premotor areas (Akkal et al., 2007; Schell and Strick, 1984) but the demonstration of projections to the primary motor cortex (Hoover and Strick, 1999), prefrontal, cingulate and orbitofrontal cortices (Middleton and Strick, 1994), regions of the inferotemporal cortex and even of posterior parietal cortex (Clower et al., 2005) enlarges the sphere of influence of the basal ganglia. Additional subcortical efferents include lateral habenula, superior colliculus, and pedunculopontine tegmental nucleus, involved in attention, locomotion, and eye movements.

Functioning of the neostriatum depends on dopamine modulation. The substantia nigra pars compacta provides dopaminergic innervation to the dorsal striatum while the ventral tegmental area provides fibers to the ventral striatum. Dopamine fibers act on both the direct pathway and indirect pathway, respectively by way of D1 and D2 dopamine receptors. The indirect system is normally the active pathway. Dopamine release activates the direct pathway. It has an inhibitory action on striatal neurons which become more sensitive to excitatory signals from the cerebral cortex, promoting thereby motor activity (Clark et al., 2005; Lalonde and Strazielle, 2010).

5. The Cerebellum

The cerebellum is classically defined as a main control structure of motor functions more specifically involved in coordination, balance, and gait. However, accumulating evidence suggests that it also plays a role in affective and higher cognitive functions, particularly in visuospatial functions (Baillieux et al., 2008; Lalonde and Botez, 1990; Schmahmann, 1997). The cerebellum is made up of a cortex segmented into lobules and three paired deep cerebellar nuclei situated from medial to lateral regions: the fastigial, the globose and emboliform grouped as the interpositus in rodents, and the dentate. The cerebellum is divided into three functional zones, namely the vermis (midline portion of the cerebellar cortex) associated with the fastigial nucleus, the paravermial region associated with the interpositus nucleus, and the hemispheric region disposed laterally on the two sides (two lateral lobes of the cerebellar cortex) associated with the dentate nucleus. Inside the cerebellum, information follows a well-organized pathway from afferent mossy or climbing fibers rising respectively indirectly or directly on Purkinje cells to the deep cerebellar nuclei, whose axons form the efferent bundles coming out of the structure (Ito, 1984).

From its role in various functional systems the cerebellum displays several afferent and efferent pathways with spinal, trigeminal, and vestibular systems, the reticular formation of the brainstem, the red nucleus, the tectum, and the hypothalamus. It also receives monoaminergic projections from the locus coeruleus, the substantia nigra, and the raphe nuclei. However, its major connections are with the cerebral cortex (Ito, 1984; Rammani, 2012). Like the cortico-striato-cortical loops, strong interconnections exist between the cerebral cortex and the cerebellum, passing through a relay in pontine nuclei in the feedforward direction to the cerebellum and successive relays in the red and thalamic nuclei in the feedback direction to the cerebral cortex. The well-known circuit concerns motor coordination in which the cerebellum receives sensory inputs from the periphery as well as from frontal, parietal, temporal, and occipital cortices and sends in turn reciprocal projections back to the primary motor cortex after a relay in the ventral lateral nucleus of the thalamus (Allen and Tsukahara, 1974). However, recent hodological studies in both primates and rodents demonstrate that the efferents from the cerebellar nuclei project to several subdivisions of the thalamus and have multiple neocortical targets not only to motor cortex, premotor cortex, and frontal eye field (Akkal et al., 2007; Hoover and Strick, 1999) but also to areas outside of the motor-related cortices, including the dorsolateral part of the prefrontal cortex (Heyder et al., 2004; Middleton and Strick, 1994) and multimodal regions in the temporal and posterior parietal lobes (Clower et al., 2005).

It is interesting to note that these anatomical findings underline the fact that the basal ganglia and the cerebellum share together the same thalamic and cortical targets. Additional experiments using retrograde transneuronal transport of viruses from injections into the neostriatum and the cerebellar cortex revealed di-synaptic pathways between these two structures (Bostan et al., 2013; Ichinohe et al., 2000). Taken together, the studies indicate that, rather than two circuits functioning in parallel, the basal ganglia and cerebellum may be components of a large functional network with the cerebral cortex involved in the higher processes of motor and cognitive functions.

In addition, physiological observations pointing towards indirect ascending fastigial projections to hippocampus, septum, and amygdala in the rat (Heath et al., 1978; Snider and Maiti, 1976) as well as corticopontine projections from the cingulate cortex and parahippocampal regions (Schmahmann and Pandya, 1993; Vilensky and Van Hoesen, 1981) indicate connections between the cerebellum and multiple limbic areas.

6. The Thalamus

Although thalamic nuclei involved in the different spatial networks have already been reported in the previous paragraphs, their presentation throughout the thalamic nuclear organization with a summary of their functional neuroanatomical features provide additional knowledge.

The thalamus is a voluminous diencephalic structure made up of a number of nuclei, which can be grouped into (1) relay nuclei for specific regions of the cerebral cortex with which they receive reciprocal connections, and (2) diffuse-projection nuclei characterized by widespread connections with the cortex and interactions with other thalamic nuclei (Groenewegen and Witter, 2004). The specific nuclei are shared in four functional groups: sensory, limbic, association, and motor thalamus for which several nuclei are relevant in spatial cognitive function.

The pertinent sensory thalamic nuclei are the lateral geniculate and the pulvinar already studied in detail in the section on the visual system. In addition, medial geniculate and ventral posterior nuclei of the thalamus relay auditory and somesthetic information respectively to superior temporal and parietal gyri (Clark, 2005; Groenewegen and Witter, 2004).

The anterior thalamic nuclei, made up of anterodorsal, anteroventral, and anteromedial nuclei in association with the lateral dorsal thalamus nucleus, constitute the limbic thalamus

(Bentivoglio et al., 1993). Connections of the different nuclei in the anterior thalamus have been related in detail in the recent article by Jankowski and colleagues (2013). The main afferents arise from the mammillary bodies by way of the mammillothalamic tract, the anterior cingulate and retrosplenial cortices, the secondary motor cortex as well as the subiculum and parahippocampal region. The main efferents project to the same regions in addition to hippocampus, associative visual areas, and prefrontal cortices (for references, see Jankowski et al., 2013). We have seen that the anterior thalamic nuclei are a central component of the Papez circuit engaged in learning and memory (Shah et al., 2012; Vertes, 2001). According to Aggleton and colleagues (2010), the anterior thalamus, in its hippocampal-diencephalic network, may be functionally organized into three parallel pathways with: (1) the anteromedial nucleus projecting to the anterior cingulate and prefrontal cortex for higher cognitive and executive functions, (2) the anterodorsal nucleus of the thalamus rich in large «head direction cells», as in the subiculum, mammillary bodies, and retrosplenial cortex for spatial orientation and navigation (Taube, 2007), (3) the anteroventral thalamic nucleus, also projecting to the retrosplenial cortex, rich in rhythmic theta activity for propagating theta signals through the Papez circuit to favor mnemonic functions (Vertes et al., 2001). The associated lateral dorsal thalamic nucleus preferentially targets the retrosplenial cortex by relaying visual information from the lateral geniculate body and superior colliculus (Bentivoglio et al., 1993).

The ventral thalamic nuclei take part in the motor circuitries involved in movement planning and control; ventral anterior and ventral lateral nuclei relay signals, respectively, from basal ganglia (medial pallidum and substantia nigra pars reticulata) (Sakai et al., 1998), and cerebellum (Aumann et al., 1994) to premotor, supplementary (Akkal et al., 2007; Sakai et al., 1999), and primary motor cortices (Hoover and Strick, 1999), and in a minor proportion to cingulate gyrus as well as to the multimodal associative areas of the parieto-temporal and frontal lobes (Aldes, 1988; Bates and Goldman-Rakic, 1993). The ventral thalamic nuclei receive, in return, modulatory inputs from frontal and parietal cortical areas (Groenewegen and Witter, 2004). The ventral medial nucleus is also considered as part of the motor thalamus. It receives the same inputs as those cited above in addition to various midbrain inputs from superior colliculus, locus coeruleus, raphe nuclei, and laterodorsal tegmental nucleus. It differs from the other ventral thalamic nuclei insofar as it has widespread projections to the frontal lobule, particularly into prefrontal areas, supposing its implication in attentional mechanisms (Herkenham, 1979; Groenewegen and Witter, 2004).

The major associative thalamic nucleus is the mediodorsal one, known to relay limbic system information to associative areas of the frontal cortex. It has strong reciprocal connections with the prefrontal cortex and frontal eye field as well as the cingulate cortex (Ray and Price, 1993). It receives additional inputs from a wide variety of subcortical sources, the amygdala, the hypothalamus, the nucleus accumbens and ventral pallidum, the substantia nigra, the superior colliculus, specially its deep layers, the laterodorsal tegmental nucleus, and the cerebellum (Groenewegen and Witter, 2004). In regard to its strong interconnections with the prefrontal cortex, recent neurophysiological studies demonstrate that the mediodorsal thalamic nucleus may participate actively in spatial working memory processes and for transforming retrospective information into prospective information (Funahashi, 2013).

Besides the specific nuclei of the thalamus, the diffuse-projecting nuclei assure a rostral continuation of the brainstem reticular formation. The intralaminar thalamic nuclei have widespread cortical connections, including the multimodal associative areas of the temporal and parietal lobes, the prefrontal cortex, the cingulate gyrus, and the primary motor cortex. The midline thalamic nuclei, composed of the reuniens and rhomboid nuclei, are the major source of excitatory thalamic projections to the hippocampus and has dense bilateral connections with the prefrontal cortex, particularly its medial part (Vertes et al., 2002; 2006) as well as anterior cingulate cortex.

Recent data even showed that a subpopulation of reuniens neurons has collateral projections to both medial prefrontal cortex and hippocampus (Hoover and Vertes, 2012). Besides their role in cortical arousal and awareness, they are suggested to be involved in long-term memory consolidation (Loureiro et al., 2012; van der Werf et al., 2002).

CONCLUSION

Lesions in animals or functional brain imaging in humans provide the opportunity to explore the neuroanatomical substrate of spatial orientation. The regions involved are numerous and were not presented exhaustively in this chapter, notably regions of the brainstem. Moreover, we did not approach certain factors like emotions or aging for example, which obviously interfere in spatial function. However, although the network is large and complex because of the multiplicity of interconnections between the different structures, some specific regions are essential and can be pointed out and organized in functional systems (Figure 2) as proposed by Kravitz et al. (2011).

The first essential structure is the posterior parietal region from which originates spatial information constructed as spatial coordinate systems to represent the personal environment (egocentric maps). The region must be considered as a central region of spatial functions because of its large two-ways projections to numerous cortical association areas in charge of cognitive functions.

The ability to navigate and learn routes by constructing extensive mental maps of the environment (allocentric maps) is mediated by the hippocampal formation in association with the parahippocampal region and the posterior cingulate cortex.

The frontal cortex provides the efferent part of spatial functions. Spatial attention and spatial working memorization are mediated by the prefrontal region while spatial guidance of motor behavior is mediated by the premotor areas. Finally, the basal ganglia and cerebellum interact with higher order structures to optimize their spatial and motor information.

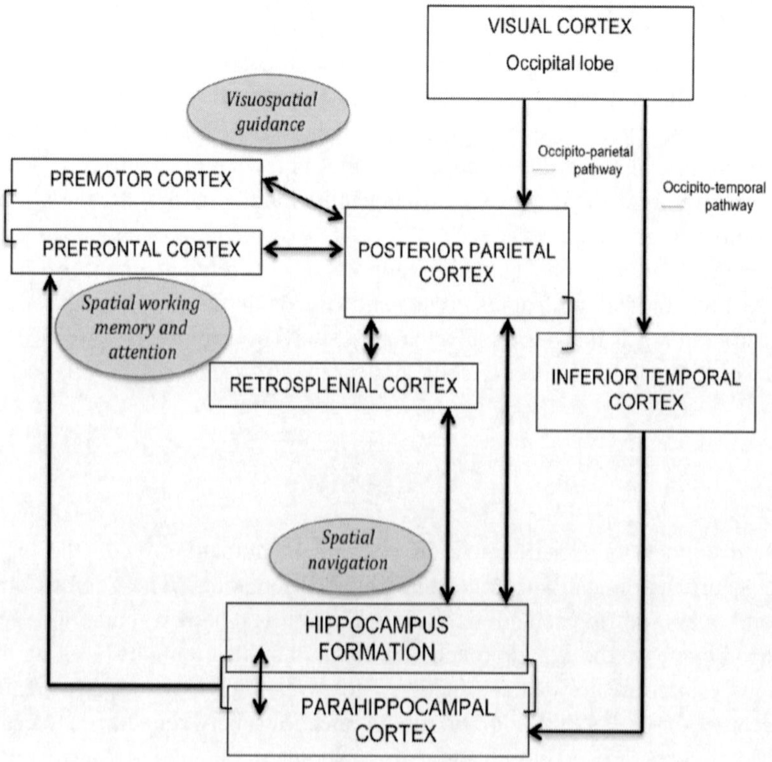

Figure 2. Schematic diagram summarizing the major structures involved in spatial orientation, grouped in the three structuro-functional systems proposed by Kravitz and colleagues (2011).

REFERENCES

Aggleton, JP; Brown, MW. Episodic memory, amnesia, and the hippocampal-anterior thalamic axis. *Behav. Brain Sci.*, 1999, 22, 425-44.

Aggleton, JP; O'Mara, SM; Vann, SD; Wright, NF; Tsanov, M; Erichsen, JT. Hippocampal-anterior thalamic pathways for memory: Uncovering a network of direct and indirect actions. *Eur. J. Neurosci.,* 2010, 31, 2292-2307.

Aguire, GK; D'Esposito, M. Topographical disorientation: A synthesis and taxonomy. *Brain,* 1999, 122, 1613-28.

Akkal, D; Dum, RP; Strick, PL. Supplementary motor area and presupplementary motor area: Targets of basal ganglia and cerebellar output. *J. Neurosci.,* 2007, 27, 10659-73.

Aldes, LD. Thalamic connectivity of rat somatomotor cortex. *Brain Res. Bull.,* 1988, 20, 333-48.

Alexander, GE; Crutcher, MD. Functional architecture of basal ganglia circuits neural substrates of parallel processing. *Trends Neurosci.,* 1990, 13, 266-71.

Allen, GI; Tsukahara, N. Cerebrocerebellar communication systems. *Physiol. Rev.,* 1974, 54, 957-1006.

Amaral, DG. A Golgi study of the cell types in the hilar region of the hippocampus in the rat. *J. Comp. Neurol.,* 1978, 182, 851-914.

Amaral, DG; Witter, MP. Hippocampal formation. In: G Paxinos (ed) *The Rat nervous system.* San Diego: Elsevier Academic Press, pp 443-93, 1995.

Andersen, RA; Cui, H. Intention, action planning, and decision making in parietal-frontal circuits. *Neuron,* 2009, 63, 568-83.

Aumann, TD; Rawson, JA; Finkelstein, DI; Horne MK. Projections from the lateral and interposed cerebellar nuclei to the thalamus of the rat: A light and electron microscopic study using single and double anterograde labelling. *J. Comp. Neurol.,* 1994, 349, 165-81.

Avillac, M; Ben Hamed, S; Duhamel, JR. Multisensory integration in the ventral intraparietal area of the macaque monkey. *J. Neurosci.,* 2007, 27, 1922-32.

Baillieux, H; De Smet, HJ; Paquier, PF; de Deyn, PP; Marië, P. Cerebellar neurocognition: Insights into the bottom of the brain. *Clin. Neurol. Neurosurg.,* 2008, 110, 763-73.

Baleydier, C; Mauguiere, F. Network organization of the connectivity between parietal area 7, posterior cingulate cortex and medial pulvinar nucleus : A double fluorescence tracer study in monkey. *Exp. Brain Res.,* 1987, 66, 365-93.

Bates, JF; Goldman-Rakic, PS. Prefrontal connections of medial motor areas in the rhesus monkey. *J. Comp. Neurol.,* 1993, 355, 211-28.

Bentivoglio, M; Kultas-Llinsky, D; Llinsky, I. Limbic thalamus: Structure, intrinsic organization, and connections. In: BA Vogt, M Gabriel (eds) *Neurobiology of cingulate cortex and limbic thalamus. A comprehensive handbook*. Boston: Birkhauser, pp 71-122, 1993.

Bisley, JW; Goldberg, ME. Neural activity in the lateral intraparietal area and spatial attention. *Science,* 2003, 299, 81-6.

Bostan, AC; Dum, RP; Strick, PL. Cerebellar networks with the cerebral cortex and basal ganglia. *Trends Cogn. Sci.,* 2013, 17, 241-54.

Bremmer, F. Multisensory space: From eye-movements to self-motion. *J. Physiol.,* 2011, 589, 815-23.

Brun, VH; Otnaess, MK; Molden, S; Steffenach, H-A; Witter, MP; Moser, M-B; Moser E. Place cells and place recognition maintained by direct entorhinal-hippocampal circuitry. *Science,* 2002, 296, 2243-46.

Buttner-Ennever, JA. Neuroanatomy of the oculomotor system. (*Reviews of oculomotor research,* vol 2*).* New-York: Elesevier, 1988.

Buzsaki, G; Chen, LS; Gage, FH. Spatial organization of physiological activity in the hippocampus regions: Relevance to memory formation. In: J Storm-Mathiesen, R Zimmer, O Ottersen (eds) Understanding Brain through the Hippocampus. *(Prog. Brain Res.,* vol 83). pp 257-68, 1990.

Caminiti, R; Ferranina, S; Johnson, PB. The sources of visual information to the primate frontal lobe: A novel role for the superior parietal lobule. *Cereb. Cortex,* 1996, 6, 319-28.

Cavada, C; Goldman-Rakic, PS. Posterior parietal cortex in rhesus monkey: I. Parcellation of areas based on distinctive limbic and sensory corticocortical connections. *J. Comp. Neurol.,* 1989, 287, 393-21.

Clark, DL; Boutros, NN; Mendez, MF. *The brain and behavior. An introduction to behavioral neuroanatomy.* Cambridge: University Press, 2005.

Clower, DM; Dum, RP; Strick, PL. Basal ganglia and cerebellar inputs to AIP. *Cereb. Cortex,* 2005, 15, 913-20.

Clower, DM; West, RA; Lynch, JC; Strick PL. The inferior parietal lobule is the target of output from the superior colliculus, hippocampus, and cerebellum. *J. Neurosci.*, 2001, 21, 6283-91.

Colby, CL; Goldberg, ME. Space and attention in parietal cortex. *Annu. Rev. Neurosci.,* 1999, 22, 319-49.

Daedwyller, SA; West, JR; Cotman, CW; Lynch, G. Physiological studies of the reciprocal connections between the hippocampus and the entorhinal cortex. *Exp. Neurol.,* 1975, 49, 35-57.

Deiana, S; Platt, B; Riedel, G. The cholinergic system and spatial learning. *Behav. Brain Res.,* 2011, 221, 389-411.

Dwyer, JA; Ingram, ML; Snow, AC; Thorp, CM; Martin, GM; Skinner, DM. The effects of bilateral lesions to the dorsal tegmental nucleus on spatial learning in rats. *Behav. Neurosci.,* 2013, 127, 867-77.

Elfgren, C; Gustafson, L; Vestberg, S; Risberg, J; Rosen, I; Ryding, E; Passant, U. Subjective experience of memory deficits related to clinical and neuroimaging findings. *Dement. Geriatr. Cogn. Disord.,* 2003, 16, 84-92.

Everitt, BJ; Robbins, TW. Central cholinergic systems and cognition. *Annu. Rev. Psychol.,* 1997, 48, 649-84.

Fogassi, L; Luppino, G. Motor functions of the parietal lobe. *Curr. Opin. Neurobiol.,* 2005, 15, 626-31.

Funahashi, S. Thalamic mediodorsal nucleus and its participation in spatial working memory processes: comparison with the prefrontal cortex. *Front. Syst. Neurosci.,* 2013, 7, article 36, 1-10.

Good, M. Spatial memory and hippocampal function: Where are we now? *Psicológica,* 2002, 23, 109-38.

Goodale, MA; Milner, AD. Separate visual pathways for perception and action. *Trends Neurosci.,* 1992,15, 20-5.

Grafton, ST; Mazziotta, JC; Presty, S; Friston, KJ; Frackowiak, RSJ; Phelps, ME. Functional anatomy of human procedural learning determined with regional cerebral blood flow and PET. *J. Neurosci.,* 1992, 12, 2542-8.

Graybiel, AM. The basal ganglia and cognitive patterns generators. *Schizophr. Bull.,* 1997, 23, 459-69.

Groenewegen, HJ; Witter, MP. The thalamus. In: G Paxinos (ed) *The rat nervous system.* San Diego: Elsevier-Academic Press, pp 408-53, 2004.

Guipponi, O; Wardak, C; Ibarrola, D; Comte, JC; Sappey-Marinier, D; Pinède, S; Ben Hamed, S. Multimodal convergence within the intraparietal sulcus of the macaque monkey. *J. Neurosci.,* 2013, 33, 4128-39.

Haber, SN; Fudge JL. The primate substantia nigra and VTA: Integrative circuitry and function. *Crit. Rev. Neurobiol.,* 1997, 11, 323-42.

Herkenham M. The afferent and efferent connections of the ventromedial thalamic nucleus in the rat. *J. Comp. Neurol.*, 1979, 183, 487-517.

Heath, RG; Dempesy, CW; Fontana, CJ; Myers, WA. Cerebellar stimulation: effects on septal region, hippocampus, and amygdala of cats and rats. *Biol. Psychiatry,* 1978, 13, 501-29.

Heyder, K; Suchan, B; Daum, I. Cortico-subcortical contributions to executive control. *Acta. Psychol.* (Amsterdam), 2004, 115, 271-89.

Hoover, JE; Strick, PL. The organization of cerebellar and basal ganglia outputs to primary motor cortex as revealed by retrograde transneuronal transport of Herpes Simplex virus type 1. *J. Neurosci.,* 1999, 19, 1446-63.

Hoover, WB; Vertes, RP. Collateral projections from nucleus reuniens of thalamus to hippocampus and medial prefrontal cortex in the rat: a single and double retrograde fluorescent labeling study. *Brain Struct. Funct.,* 2012, 217, 191-209.

Huang, C; Wahlund, L-O; Svensson, L; Winblad, B; Julin, P. Cingulate cortex hypoperfusion predicts Alzheimer's disease in mild cognitive impairment. *BMC Neurol.,* 2002, 2, 9 (doi: 10.1186/1471-2377-2-9).

Huerta, MF; Harting, JK. The mammalian superior colliculus : Studies of its morphology and connections. In: H Vanegas (ed) *Comparative neurology of the optic tectum.* New York: Plenum Press, 1984.

Iaria, G; Petrides, M; Dagher, A; Pike, B; Bohbot, VD. Cognitive strategies dependent on the hippocampus and caudate nucleus in human navigation: Variability and change with practice. *J. Neurosci.,* 2003, 23, 5945-52.

Ichinohe, N; Mori, F; Shoumara, K. A di-synaptic projection from the lateral cerebellar nucleus to the laterodorsal part of the striatum via the central lateral nucleus of the thalamus in the rat. *Brain Res.,* 2000, 880, 191-7.

Insausti, R; Herrero, MT; Witter MP. Entorhinal cortex of the rat: Cytoarchitectonic subdivisions and the origin and distribution of cortical efferents. *Hippocampus,* 1997, 7, 146-83.

Ito, M. *The cerebellum and neural control.* New York: Raven Press, 1984.

Jacobs, LF; Schenk, F. Unpacking the cognitive map: the parallel map theory of hippocampal function. *Psychol. Rev.,* 2003, 110, 285-315.

Jankowski, MM; Ronnqvist, KC; Tsanov, M; Vann, SD; Wright, NF; Erchsen, JT; Aggleton, JP; O'Mara, SM. The anterior thalamus provides a subcortical circuit supporting memory and spatial navigation. *Front. Syst. Neurosci.,* 2013, 7, article 45, 1-13.

Kaas, JH. Why does the brain have so many visual areas? *J. Cogn. Neurosci.*, 1989, 1, 121-35.

Kakei, S; Hoffman, DS; Strick, LP. Muscle and movement representation in the primary motor cortex. *Science,* 1999, 285, 2136-39.

Katsuki, F; Constandinidis, C. Unique and shared roles of the posterior parietal and dorsolateral prefrontal cortex in cognitive functions. *Front. Int. Neurosci.,* 2012, 6, article 17, 94-107.

Kobayashi, Y; Amaral, DG. Macaque monkey retrosplenial cortex: II. Cortical afferents. *J. Comp. Neurol.,* 2003, 466, 48-79.

Kolb, B; Walkey, J. Behavioural and anatomical studies of the posterior parietal cortex in the rat. *Behav. Brain Res.,* 1987, 23, 127-45.

Kravitz, DJ; Saleem, KS; Baker, CI; Mishkin, M. A new neural framework for visuospatial processing. *Nat. Rev. Neurosci.,* 2011, 12, 217-30.

Lalonde, R; Botez, MI. The cerebellum and learning processes in animals. *Brain Res. Brain Res. Rev.,* 1990, 15, 325-32.

Lalonde, R; Strazielle, C. Basal ganglia disorders in genetic models and experimentally-induced lesions. In: AV Kalueff, CL Bergner (eds) *Transgenic and mutant tools to model brain disorders.* (Neuromethods, vol 44). New York: Humana Press, pp 91-112, 2010.

Lavenex, P; Amaral, DG. Hippocampal–neocortical interaction: a hierarchy of associativity. *Hippocampus,* 2000, 10, 420–30.

Lepage, M; Habib, R; Tulving E. Hippocampal PET activations of memory encoding and retrieval: the HIPER model. *Hippocampus,* 1998, 8, 313-22.

Levitt, P; Rakic, P; Goldman-Rakic, P. Region-specific distribution of catecholamine afferents in primate cerebral cortex: a fluorescence histochemical analysis. *J. Comp. Neurol.,* 1984, 227, 23-36.

Lewis, DA; Foote, SL; Goldstein, M; Morrison, JH. The dopaminergic innervation of monkey prefrontal cortex: a tyrosine hydroxylase immunohistochemical study. *Brain Res.,* 1988, 449, 225-43.

Lopez, C; Blanke, O. The thalamocortical vestibular system in animals and humans. *Bain Res. Rev.,* 2011, 67, 119-46.

Lopes da Silva, L; Witter, FH; Boeijinga, PH; Lohman, AHM. Anatomical organization and physiology of the limbic cortex. *Physiol. Rev.,* 1990, 70, 453-511.

Loureiro, M; Cholvin, T; Lopez, J; Merienne, N; Latreche, A; Cosquer, B; Geiger, K; Kelche, C; Cassel, J-C; Pereira de Vasconcelos, A. The ventral midline thalamus (reuniens and rhomboid nuclei) contributes to the persistence of spatial memory in rats. *J. Neurosci.,* 2012, 32, 9947-59.

Maguire, AE; Burgess, N; O'Keefe, J. Human spatial navigation: cognitive maps, sexual dimorphism, and neural substrates. *Curr. Opin. Neurobiol.,* 1999, 9, 171-7.

McDonald, RJ; White, NM. Parallel information processing in the watermaze: evidence for independent memory systems involving dorsal striatum and hippocampus. *Behav. Neural Biol.,* 1994, 109, 579–93.

Mega, MS; Cummings, JC; Salloway, S; Malloy, P. The limbic system: An anatomic, phylogenetic, and clinical perspective. *J. Neuropsychiatry Clin. Neurosci.,* 1997, 9, 315-30.

Mesulam, MM; Mufson, EJ; Wainer, BH; Levey, AI. Central cholinergic pathways in the rat: an overview based on an alternative nomenclature (Ch1-Ch6). *Neuroscience,* 1983, 10, 1185–201.

Middleton, FA; Strick, PL. Anatomical evidence for cerebellar and basal ganglia involvement in higher cognitive function. *Science,* 1994, 266, 458-61.

Montague, PR; Hyman, SE; Cohen, JD. Computational roles for dopamine in behavioural control. *Nature,* 2004, 431, 760-7.

Moscovitch, M; Rosenbaum, RS; Gilboa, A; Addis, DR; Westmacott, R; Grady, C; McAndrews, MP; Levine, B; Black, S; Winocur, G; Nadel L. Functional neuroanatomy of remote episodic, semantic, and spatial memory : a unified account based on multiple trace theory. *J. Anat.,* 2005, 207, 35-66.

Moser, MB; Moser, EI. Distributed encoding and retrieval of spatial memory in the hippocampus. *J. Neurosci.,* 1998, 18, 7535-42.

Naber, PA; Lopes da Silva, FH; Witter MP. Reciprocal connections between the entorhinal cortex and hippocampal fields CA1 and the subiculum are in register with the projections from CA1 to the subiculum. *Hippocampus,* 2001, 11, 99-104.

Noudoost, B; Moore, T. Control of visual cortical signals by prefrontal dopamine. *Nature,* 2011, 474, 372-5.

O'Keefe, J; Dostrovsky, J. The hippocampus as a spatial map: Preliminary evidence from unit activity in the freely moving rat. *Brain Res.,* 1971, 34, 171-75.

Palomero-Gallagher, N; Zilles, K. Isocortex. In: G Paxinos (ed) *The rat nervous system.* San Diego: Elsevier Academic Press, pp 728-57, 2004.

Papez, JW. A proposed mechanism of emotion. *Arch. Neurol. Psychiatry,* 1937, 38, 725-43.

Parron, C; Poucet, B; Save, E. Cooperation between the hippocampus and the entorhinal cortex in spatial memory : A disconnection study. *Behav. Brain Res.,* 2006, 170, 99-109.

Picard, N; Strick, PL. Motor areas of the medial wall: A review of their location and functional activation. *Cereb. Cortex,* 1996, 6, 342-53.

Rammani N. Frontal lobe and posterior parietal contributions to the cortico-cerebellar system. *Cerebellum,* 2012, 11, 366-83.

Ray, JP; Price, JL. The organization of projections from the mediodorsal nucleus of the thalamus to orbital and medial prefrontal cortex in macaque monkeys. *J. Comp. Neurol.,* 1993, 337, 1-31.

Reep, RL; Chandler, HC; King, V; Corwin, JV. Rat posterior parietal cortex : topography of corticocortical and thalamic connections. *Exp. Brain Res.,* 1994, 100, 67-84.

Rizzolatti, G; Luppino, G; Matelli, M. The classic supplementary motor area is formed by two independent areas. In: HO Lunders (ed) *Supplementary sensorimotor area.* Philadelphia: Lippincott-Raven, pp 45-66, 1996.

Rodriguez, F; López, JC; Vargas, JP; Broglio, C; Gómez, Y; Salas, C. Spatial memory and hippocampal pallium through vertebrate evolution: Insights from reptiles and teleost fish. *Brain Res. Bull.,* 2002, 57, 499-503.

Roberts, AC; Robbins, TW; Weiskrantz, L. *The prefrontal cortex. Executive and cognitive functions.* New York: Oxford University Press, 2000.

Sahyoun, C; Floyer-Lea, A; Johansen-Berg, H; Matthews, PM. Towards an understanding of gait control: brain activation during the anticipation, preparation and execution of foot movements. *NeuroImage,* 2003, 21, 568-75.

Sakai, ST; Grofova, I; Bruce, K. Nigrothalamic projections and nigrothalamocortical pathway to the medial agranular cortex in the rat: single- and double-labeling light and electron microscopic studies. *J. Comp. Neurol.,* 1998, 391, 506-25.

Sakai, ST; Inase, M; Tanji, J. Pallidal and cerebellar inputs to thalamo-cortical neurons projecting to the supplementary motor area in *Macaca Fuscata*: A triple-labeling light microscopic study. *Anat. Embryol. (Berl),* 1999, 199, 9-19.

Sakata, H; Taira, M; Kuisunoki, M; Murata, A; Tanaka, Y. The parietal association cortex in depth perception and visual control of hand action. *Trends Neurosci.,* 1997, 20, 350-7.

Sawaguchi, T; Goldman-Rakic, PS. D1 dopamine receptors in prefrontal cortex: involvement in working memory. *Science,* 1991, 251, 947-50.

Schell, GR; Strick, PL. The origin of thalamic inputs to the arcuate premotor and supplementary premotor areas. *J. Neurosci.,* 1984, 4, 539-60.

Schliebs, R; Arendt, T. The cholinergic system in ageing and neural degeneration. *Behav. Brain Res.,* 2011, 221, 555-563.

Schmahmann, JD. *The cerebellum and cognition.* San Diego: Academic Press, 1997.

Schmahmann, JD; Pandya, DN. Anatomic organization of the basilar pontine projections from prefrontal cortices in rhesus monkey. *J. Neurosci.,* 1997, 17, 438-58.

Schmahmann, JD; Pandya, DN. Prelunate, occipitotemporal, and parahippocampal projections to the basis pontis in rhesus monkey. *J. Comp. Neurol.,* 1993, 337, 94-112.

Selemon, LD; Goldman-Rakic, PS. Common cortical and subcortical targets of the dorsolateral prefrontal and posterior parietal cortices in the rhesus monkey: Evidence for a distributed neural network subserving spatially guided behavior. *J. Neurosci.,* 1988, 8, 4049-68.

Shah, A; Jhawar, SS; Goel, A. Analysis of the anatomy of the Papez circuit and adjoining limbic system by fiber dissection techniques. *J. Clin. Neurosci.,* 2012, 19, 289-298.

Shibata, H. Efferent projections from the anterior thalamic nuclei to the cingulate cortex in the rat. *J. Comp. Neurol.,* 1993, 330, 533-42.

Silva, AJ; Giese, KP; Fedorov, NB; Frankland, PW; Kogan, JH. Molecular, cellular, and neuroanatomical substrates of place learning. *Neurobiol. Learn. Mem.,* 1998, 70, 44-61.

Snider, RS; Maiti, A. Cerebellar contributions to the Papez circuit. *J. Neurosci. Res.,* 1976, 2, 133-46.

Squire, LR. Memory and the hippocampus: a synthesis from findings with rats, monkeys, and humans. *Psychol. Rev.,* 1992, 99, 195-231.

Sommer, MA; Wurtz, RH. Frontal eye field neurons orthodromically activated from the superior colliculus. *J. Neurophysiol.,* 1998, 80, 3331-5.

Steckler, T; Inglis, W; Winn, P; Sahgal, A. The pedunculopontine tegmental nucleus: a role in cognitive processes? *Brain Res. Brain Res. Rev.,* 1994, 19, 298-318.

Stuss, DT; Benson, DF. *The Frontal Lobes*. New York: Raven Press, 1986.

Sutherland, RJ; Hoesing, JM. Posterior cingulate cortex and spatial memory: A microlimnology analysis. In: BA Vogt, M Gabriel (eds) *Neurobiology of cingulate cortex and limbic thalamus. A comprehensive handbook*. Boston: Birkhauser, pp 461-77, 1993.

Taube, JS. The head direction signal: Origins and sensori-motor integration. *Annu. Rev. Neurosci.,* 2007, 30, 181-207.

Taube, JS; Muller, RU; Ranck, JBJr. Head-direction cells recorded from the post-subiculum in freely moving rats. I. Description and quantitative analysis. *J. Neurosci.,* 1990, 10, 420-35.

Ungerleider, LG; Mishkin, M. Two cortical visual systems. In: DJ Ingle, MA Goodale, RJW Mansfield (eds) *Analysis of visual behavior.* Cambridge, MA: MIT Press, pp 549-86, 1982.

Van der Werf, YD; Witter, MP; Groenewegen, HJ. The intralaminar and midline nuclei of the thalamus. Anatomical and functional evidence for participation in processes of arousal and awareness. *Brain Res. Brain Res. Rev.,* 2002, 39, 107–40.

Van Groen, T; Lopes da Silva, FH. The organization of the reciprocal connections between the subiculum and the entorhinal cortex in the rat: II. An electrophysiological study. *J. Comp. Neurol.,* 1986, 251, 111-20.

Van Hoesen, GW. The parahippocampal gyrus. New observations regarding its cortical connections in the monkey. *Trends Neurosci.,* 1982, 5, 345-50.

Vertes, RP. Analysis of projections from the medial prefrontal cortex to the thalamus in the rat, with emphasis on nucleus reuniens. *J. Comp. Neurol.*, 2002, 442, 163-87.

Vertes, RP; Albo, Z; Viana Di Prisco, G. Theta-rhythmically firing neurons in the anterior thalamus: Implications for mnemonic functions of Papez's circuit. *Neuroscience,* 2001, 104, 619-25.

Vertes, RP; Hoover, WB; Do Valle, AC; Sherman, A; Rodriguez JJ. Efferent projections of reuniens and rhomboid nuclei of the thalamus in the Rat. *J. Comp. Neurol.,* 2006, 499, 768-96.

Vilensky, JA; Van Hoesen, GW. Corticopontine projections from the cingulate cortex in the rhesus monkey. *Brain Res.,* 1981, 205, 391-5.

Walsh, TJ; Herzog, CD; Gandhi, C; Stackman, RW; Wiley, RG. Injection of IgG 192-saporin into the medial septum produces cholinergic hypofunction and dose dependent working memory deficits. *Brain Res.,* 1996, 726, 69–79.

Winn, P; Brown, VJ; Inglis, WL. On the relationships between the striatum and the pedunculopontine nucleus. *Crit. Rev. Neurobiol.,* 1997, 11, 241-61.

Wise, SP; Boussaloud, D; Johnson, PB; Caminiti, R. Premotor and parietal cortex: corticocortical connectivity and combinatorial computations. *Annu. Rev. Neurosci.,* 1997, 20, 25-42.

Witter, MP; Groenewegen, HJ; Lopes da Silva, FH; Lohman, AHM. Functional organization of the extrinsic and intrinsic circuitry of the parahippocampal region. *Prog. Neurobiol.,* 1989, 33, 161-253.

Witter, MP; Naber, PA; van Haeften, T; Machielsen, WC; Rombouts, SA; Barkhof, F; Scheltens, P; Lopes da Silva, FH. Cortico-hippocampal communication by way of parallel parahippocampal-subicular pathways. *Hippocampus,* 2000, 10, 398–410.

Yantis, S; Schwarzbach, J; Serences, JT; Carlson, RL; Steinmetz, MA; Pekar, JJ; Courtney, SM. Transient neural activity in human parietal cortex during spatial attention shifts. *Nature Neurosci.*, 2002, 5, 995-1002.

Zeki, S. *A vision of the brain.* Oxford: Blackwell Scientific Publications, 1993.

Zola-Morgan, SM; Squire, LR. Neuroanatomy of memory. *Annu. Rev. Neurosci.,* 1993, 16, 547-63.

In: Spatial Orientation in Humans and Animals
Editor: Robert Lalonde
ISBN: 978-1-63463-001-6
© 2014 Nova Science Publishers, Inc.

Chapter 4

SPATIAL DEFICITS IN PATIENTS WITH ALZHEIMER'S DISEASE

Robert Lalonde*
Université de Rouen, Dépt. Psychologie,
Mont-Saint-Aignan Cedex France

1. BIOLOGICAL CHARACTERISTICS OF ALZHEIMER'S DISEASE

1.1. Neuropathology

Definite Alzheimer's disease is diagnosed after post-mortem evidence of extracellular neuritic plaques and intracellular neurofibrillary tangles in the brain (Perl, 2000). Another neuropathologic sign comprises granulovacuolar degeneration marked by intracytoplasmic vesicles (Su et al., 2002). Neuritic plaques contain A-beta-protein (Aβ) and cellular debris, Aβ-positive neurofibrillary tangles contain hyperphosphorylated tau (Perl, 2000; Perry et al., 1992). Cored plaques are distinguishable from diffuse ones by their β-pleated structure and later onset (Yamaguchi et al., 1988; Thal et al., 2000). Diffuse plaques contain only $A\beta_{42}$ but mature plaques both $A\beta_{42}$ and $A\beta_{40}$ (Iwatsubo et al., 1994). Aβ plaques are surrounded by activated microglia and astrocytes as well as inflammation-related proteins (Eikelenboom et al., 2002).

In comparison to age-matched controls, grey matter density was lower in patients with Alzheimer's disease at the level of neocortex, hippocampus, amygdala, and insula (Frisoni et al., 2002). Likewise, grey matter volume decreased in neocortical and medial temporal regions (Double et al., 1996; Karas et al., 2004), extending to white matter in neocortex (Bartzokis et al., 2003) and hippocampus (Callen et al., 2001; Jack et al., 1998; Laakso et al., 1998). The white matter loss is partly caused by dying back phenomena from grey matter loss but also by plaque-induced demyelination (Mitew et al., 2010). Entire limbic pathways may be affected by volume shrinkage, including parahippocampal cortex, septum, anterior thalamus, mamillary bodies, posterior cingulate, and orbitofrontal cortex (Callen et al., 2001),

* Email: robert.lalonde@univ-rouen.fr

as well as entorhinal and perirhinal cortices (Juottonen et al., 1998). The volume loss appeared first in medial temporal and parietal lobes, progressed to infero-lateral temporal neocortex, and then spread to prefrontal cortex (Scahill et al., 2002). Signs of atrophy appeared on magnetic resonance imaging (MRI) scans even in presymptomatic individuals with autosomal dominant Alzheimer's disease (Fox et al., 2001) or those with mild cognitive impairment (MCI), a presymptomatic phase towards dementia (Duara et al., 2013). Alzheimer-related atrophy extends to subcortical structures innervating neocortex, such as basal nucleus of Meynert, dorsal raphe, locus coeruleus, and to a lesser extent substantia nigra pars compacta, sources of acetylcholine, 5-hydroxytryptamine (5HT), noradrenaline, and dopamine, respectively (Lyness et al., 2003; Rogers et al., 1985), leading to lower levels of these neurotransmitters in the neocortex (Coyle et al., 1983; Palmer et al., 1987; Whitford, 1986). The thalamus was also reduced in volume (de Jong et al., 2008). Moreover, grey matter loss was observed in caudate nucleus (Rombouts et al., 2000) and vermal cerebellum (Sjöbeck and Englund, 2001) and total volume loss was observed in putamen (de Jong et al., 2008), all three regions with some diffuse but few neuritic Aβ plaques (Moran et al., 1993). As predicted by the amyloid cascade hypothesis (Hardy, and Selkoe, 2002), the rate of atrophy was correlated with the uptake of Pittsburgh Compound-B (^{11}PIB) (Archer et al., 2006), a molecule that binds to insoluble Aβ plaques on positron emission tomography (PET) scans (Herholz and Ebmeier, 2011).

Intracellular $A\beta_{42}$ accumulation in multivesicular bodies at pre- and postsynaptic sites (Takahashi et al., 2002) as well as the cell membrane (Yamaguchi et al., 2000) probably plays an active role in neurodegeneration (Gouras et al., 2000), in that non-fibrillar Aβ induced apoptotic cell death in rat cortical neuronal cultures (Pillot et al., 1999). Although DNA fragmentation is observed in Alzheimer brain, other morphological markers of apoptosis (chromatin condensation, nuclear blebbing, and apoptotic bodies) are rare (Anderson et al., 2000; Broe et al., 2001; Colurso et al., 2003; Lassmann et al., 1995; Troncoso et al., 1996), so that this condition is more rightly called "abortosis". In certain cells manifesting abortosis, upstream but not downstream caspases were activated (Raina et al., 2001), although one of the latter (caspase-3) was activated in cells with granulovacuolar degeneration (Su et al., 2002). Despite severe cholinergic cell loss, DNA fragmentation was not observed in the basal nucleus of Meynert (Wu et al., 2005).

1.2. Genetics

Mutations of *APP*, *PS1,* and *PS2* genes (Campion et al., 1999; De Jonghe and Van Broeckhoven, 1999; Larner and Doran, 2006; Tanzi and Bertram, 2001) along with duplication at the *APP* locus (Rovelet-Lecrux et al., 2006) cause autosomal dominant Alzheimer's disease. For some *PS1* mutations, dementia is diagnosed as early as 40 years of age (Campion et al., 1996; De Jonghe and Van Broeckhoven, 1999). Despite similar neuropathological (Mendez et al., 1992) and neuropsychological (Swearer et al., 1992) markers, brain atrophy (Gregory et al., 2006) and metabolism (Mosconi et al., 2003) were more severely affected in autosomal dominant than in sporadic cases. Patients with Down's syndrome develop the same neuropathological lesions due to an extra copy of *APP* located on chromosome 21 (Mann and Esiri, 1989).

APP encodes beta-amyloid precursor protein, from which derive $A\beta_{42}$ or $A\beta_{40}$ after successive proteolytic cleavages catalyzed by β- and γ-secretases at N- and C-terminals, respectively. In Alzheimer temporal cortex, β-secretase activity declined while α-secretase activity associated with the non-amyloid pathway rose (Tyler et al., 2002). The APP_{695} isoform is mostly found in neurons, while APP_{751} and APP_{770} isoforms are found in glia (Rohan de Silva et al., 1997). PS1 and PS2 consist of catalytic subunits of γ-secretase (Wolfe et al., 1999). In addition to Aβ plaques, Alzheimer-related pathologies may be directly caused by APP (Neve et al., 2000) or its C-terminal fragments (Turner et al., 2003).

1.3. Cholinergic Enzymes

Post-mortem analyses of Alzheimer cortex reveal decreased activity of choline acetyltransferase (ChAT), the acetylcholine-synthesizing enzyme (Coyle et al., 1983; Efange et al., 1997; Gil-Bea et al., 2005; Hardy et al., 1985; Minger et al., 2001) as well as decreased binding of the vesicular acetylcholine transporter (Efange et al., 1997), at least partly due to basal forebrain neurodegeneration (Coyle et al., 1983; Hardy et al., 1985; Lyness et al., 2003; Wu et al., 2005). As seen with ChAT, the activity of acetylcholinesterase (AChE), the non-specific catabolic enzyme, decreased in post-mortem samples (Blusztajn and Berse, 2000; Coyle et al., 1983; Gil-Bea et al., 2005; Hardy et al., 1985), in particular the tetrameric isoform related to enzymatic activity (Atack et al., 1983; Sberna et al., 1998), as well as *in vivo* on PET scans (Herholz et al., 2004; Kuhl et al., 1999; Okamura et al., 2008; Rinne et al., 2003). The enzymatic changes reflect reduced cholinergic activity in parenchyma but also probably on blood vessels, because muscarinic and nicotinic receptors are found on blood vessels affecting blood flow throughout the brain (Van Beek and Claassen, 2011). However, AChE staining is sometimes enhanced in hippocampal subregions (Hyman et al., 1987), either as a compensatory response or because of its link with Aβ plaques (Kalaria et al., 1992; Moran et al., 1993), thereby playing a role in disease progression (Talesa, 2001) via effects on neurite growth promotion, synaptogenesis, cell adhesion, and neurotransmitter release (Soreq and Seidman, 2001). In particular, AChE potentiated Aβ aggregation and neurotoxicity and induced nitric oxide synthase (iNOS) expression *in vitro* (Talesa, 2001; von Bernhardi et al., 2003). In addition, Aβ formation appeared earlier and denser after breeding APP_{695}SWE transgenic mice with *ACHE* overexpressing lines (Rees et al., 2003). Moreover, release of AchE from Aβ plaques in transgenic mice diminished Aβ aggregates and ameliorated spatial learning (Dinamarca et al., 2008).

2. Neuropsychological Characteristics

2.1. General

During the patient's lifetime, probable Alzheimer's disease is diagnosed when progressive loss occurs in remembering new items (anterograde amnesia) along with a deficit in at least one of four instrumental functions: language (aphasia), object use (apraxia), form recognition (agnosia), or step-by-step planning, as first defined by McKhann et al. (1984).

Several dementia rating scales possess a high degree of concurrent validity of the disease, notably the Mini-mental state examination (Bracco et al., 1990; Giordani et al., 1990; Kokmen et al., 1991). Memory and verbal fluency deficits are usually the initial symptoms (Amieva et al., 2005). Progressive amnesia is sometimes the only sign observed over a period of several years (Weintraub and Mesulam, 1993). A loss in factual information (declarative memory) is found for verbal and spatial items (Amieva et al., 2005; Graham et al., 2004; Helmes and Ostbye 2002; Jacobs et al., 1995; Kessels et al., 2005; Lee et al., 2003), whereas, except for delay and trace conditioning of the eyeblink response (Woodruff-Pak and Papka, 1996; Woodruff-Pak et al., 1996), procedural memory is usually spared, including pursuit-rotor (Eslinger and Damasio, 1986; Jacobs et al., 1999) and verbal (Karlsson et al., 2002) or spatial (Kessels et al., 2005) priming tasks. There is also an impairment in prepulse inhibition of the acoustic startle response, which implies a deterioration in sensory gating (Ueki et al., 2006).

2.2. Brain-Behavior Relations

2.2.1. Neuropathological Markers

As predicted by the amyloid cascade hypothesis, Aβ plaque load in Alzheimer brain is often correlated with cognitive scores (Gold et al., 2001; Kanne et al., 1998; Martin et al., 1987; Näslund et al., 2000; Nielson et al., 1996; Parvathy et al., 2001; Samuel et al., 1994; Terry et al., 1991). Both diffuse and cored $A\beta_{42}$ or $A\beta_{40}$ levels were correlated with dementia severity (Parvathy et al., 2001).

Moreover, the significant correlation between either $A\beta_{40}$ or $A\beta_{42}$ concentrations and dementia scores appeared in several brain regions (Näslund et al., 2000). Dementia severity was also associated with lower $A\beta_{42}$ concentrations in cerebrospinal fluid, a consequence of brain deposition (Riemenschneider et al., 2002), as well lower serum concentrations of apoliprotein AI, an Aβ-binding protein (Merched et al., 2000).

In some reports, dementia severity was correlated with neurofibrillary tangles but not Aβ plaque load (Arriagada et al., 1992; Bierer et al., 1995) or to a lesser extent (Giannakopoulos et al., 2003, Tiraboschi et al., 2004). This discordance is attributable to the early plateau reached by Aβ plaque deposition, unlike the steady rise in neurofibrillary tangles (Morris et al., 1995; Tiraboschi et al., 2004). Indeed, soluble or insoluble $A\beta_{40}$ and $A\beta_{42}$ concentrations were not correlated with disease duration in one study (Ingelsson et al., 2004), though in another $A\beta_{42}$ but not $A\beta_{40}$ concentrations were correlated with it (Kumar-Singh et al., 2006). While soluble Aβ was correlated with neurofibrillary tangles and the patient's death, insoluble Aβ was not (McLean et al., 1999).

Neurofibrillary tangle pathology is a predictor not only of cognitive deficits (Arriagada et al., 1992, Bierer et al., 1995; Gold et al., 2001; Samuel et al., 1994; Terry et al., 1991) but also of neuropsychiatric signs such as agitation (Tekin et al., 2001). As the end result of accumulating plaques and tangles, cognitive deficits are particularly associated with synapse numbers (Minger et al., 2001; Samuel et al., 1994; Terry et al., 1991) or concentrations of synaptophysin (Samuel et al., 1994; Terry et al., 1991), a synaptic protein that declines in Alzheimer brain (Reddy et al., 2005).

2.2. Cholinergic Markers

In addition to Aβ plaques and neurofibrillary tangles, brain-behavior relations in Alzheimer patients have been demonstrated with respect to cholinergic enzymes. Indeed, poorer cognitive performances were linearly correlated with lower activities of ChAT (Baskin et al., 1999; DeKosky et al., 1992; Gibson et al., 2000; Gil-Bea et al., 2005; Minger et al., 2001; Pappas et al., 2001) and AChE (Bohnen et al., 2005; Gil-Bea et al., 2005; Shinotoh et al., 2000).

Table 1. Spatial deficits in patients with Alzheimer's disease

Function	Tests	Reference
visuoconstructive skills	block design, object assembly	Becker et al., 1988; Jeste et al., 1992; Mohr et al., 1990
visuoconstructive skills	simple drawings	Guariglia and Nitrini, 2009; Jeste et al., 1992; Morris et al., 1989
visuoconstructive skills	Rey or Benton figures	Becker et al., 1988; Brouwers et al., 1984; Mohr et al., 1990; Possin et al., 2011
map reading	Money's road-map	Flicker et al., 1988; Mapstone et al., 2003; Mohr et al., 1990; Rainville et al., 2002a
map reading	virtual town	Pengas et al., 2010
remote spatial knowledge	Fargo road-map	Cherrier and Tröster, 1998
personal route-finding	own city and country	Guariglia and Nitrini, 2009
working memory	Corsi block-tapping	Guariglia and Nitrini, 2009; Huntley et al., 2011; Possin et al., 2011; Spinnler et al., 1988; Toepper et al., 2008; Wiechmann et al., 2011
working memory	delayed response or alternation and spatial reversal	Freedman and Oscar-Berman, 1989a,b
working memory	recognition of locations	Alescio-Lautier et al., 2007; Nordhal et al., 2005
long-term memory	recall of Rey or figure reproduction of the Wechsler memory scale	Mohr et al., 1990
long-term memory	mazes	Brouwers et al., 1984; Ott et al., 2008
long-term memory	route-walking	Hort et al., 2007; Monacelli et al., 2003; Uc et al., 2004

3. Spatial Functions

As in other patient populations, care must be taken to ensure that deficits that seem to be due to spatial cognition are not the consequence of deficits in visual perception, such as the loss in contrast sensitivity reported in patients with Alzheimer's disease (Nissen et al., 1985). Despite this potentially confounding factor, especially in tasks requiring a high degree of visual acuity, Alzheimer's disease has often been characterized by impaired visuoconstructive skills and spatial reasoning as well as short-term, working, and long-term memory (Table 1).

3.1. Visuoconstructive Skills

Patients with Alzheimer's disease are susceptible to constructional apraxia (Graham et al., 2004; Helmes and Ostbye 2002; Nielson et al., 1996; Serra et al., 2014), defined as a difficulty in copying geometric figures or reproducing them with building blocks (chapter 1). In particular, Alzheimer patients had poorer scores than controls on block design (Becker et al., 1988; Jeste et al., 1992; Mohr et al., 1990) and object assembly (Mohr et al., 1990) subtests of the Wechsler intelligence scale. Block design was also impaired in subjects with MCI (Nordahl et al., 2005). In addition, Alzheimer patients were impaired while copying the Rey complex figure (Becker et al., 1988; Brouwers et al., 1984; Mohr et al., 1990) or a simplified version of it, the Benson figure (Possin et al., 2011).

Copying accuracy of the Benson figure was correlated with lower volumes in right parietal cortex, crucially involved in spatial functions. Spatial impairments were likewise observed when patients drew simpler drawings such as a circle, a diamond, a cube, intersecting rectangles (Morris et al., 1989), a house (Guariglia and Nitrini, 2009), or clocks (Jeste et al., 1992). The accuracy in drawings clocks and houses was correlated with caregiver accounts of spatial disorientation, supporting the view that these measures possess construct validity in this condition (Henderson et al., 1989). Copying accuracy of the Rey complex figure was correlated with copying embedded figures of the Mini-mental state examination, an indication of convergent validity between two commonly used tests (Giordani et al., 1990).

Several underlying causes of constructional apraxia have been implicated, notably deficits in flow perception (Kavcic et al., 2006; Tetewsky and Duffy, 1999), visual search (Parasuraman et al., 2000; Tales et al., 2004), abstract reasoning (Helmes and Ostbye 2002), and executive functions (Collette et al., 1999; Graham et al., 2004; Rainville et al., 2002b).

3.2. Map Reading

As described after damage to frontal or parietal association cortex (chapter 1), patients with Alzheimer's disease were impaired relative to controls in the Money road-map test of directional sense (Flicker et al., 1988; Mapstone et al., 2003; Mohr et al., 1990; Rainville et al., 2002a).

Moreover, Alzheimer patients were deficient in two route-finding tasks inside a virtual town observed on a computer screen (Pengas et al., 2010). The patients were also impaired in the Fargo map test of remote spatial knowledge (Cherrier and Tröster, 1998). Impairments were likewise prominent in Alzheimer patients residing in San Paolo, Brazil when asked to describe a route from his/her own house to the nearest grocery store or in estimating the location of Brazilian cities and San Paolo landmarks (Guariglia and Nitrini, 2009).

3.3. Spatial Short-Term, Working, and Long-Term Memory

Short-term and working memory are assessed respectively by forward and backward spans of the Corsi block-tapping test or the block-tapping subtest of the Wechsler memory scale-revised (chapter 1). Relative to controls, Corsi block-tapping was impaired in patients with Alzheimer's disease (Spinnler et al., 1988), even at the mild (Huntley et al., 2011) or moderate (Guariglia and Nitrini, 2009) stage of dementia. Both forward and backward spans of block-tapping was impaired in these patients (Possin et al., 2011; Toepper et al., 2008; Wiechmann et al., 2011). When spatial forward span was assessed with circles on a computer monitor, Alzheimer and MCI subjects were impaired (Maki et al., 2010). Both groups were also deficient in spatial short-term memory in the Four mountains test (Bird et al., 2010) and in recognition tasks of object location (Alescio-Lautier et al., 2007; Nordhal et al., 2005), so that these tests are considered useful in early detection of the dementia syndrome (Iachini et al., 2009).

Other spatial working memory tests include spatial delayed response and left-right alternation derived from research in animals (chapters 1 and 2). Delayed response and delayed alternation were impaired in patients with Alzheimer's disease relative to controls (Freedman and Oscar-Berman, 1989b). Left-right discrimination and reversal learning are two additional tests derived from research in animals (chapters 1 and 2). Spatial reversal learning was impaired in patients with Alzheimer's disease relative to controls, but not original training of the left-right response (Freedman and Oscar-Berman, 1989a). These results indicate that patients are especially lacking in working memory and mental flexibility. Moreover, dual task performances declined in patients when a visual tracking task was added to stimulate the visuospatial sketchpad, revealing a deterioration of the central executive (Baddeley et al., 1991).

In addition to spatial short-term or working memory, long-term memory is affected in this syndrome. Relative to controls, patients with Alzheimer's disease had impaired recall of the Rey complex figure as well as figure reproduction in the Wechsler memory scale (Mohr et al., 1990). A decline in spatial memory appears early in the course of the disease, as indicated by deficits found on delayed recall of the Rey complex figure (Marra et al., 2011) and other spatial items (Flicker et al., 1991) in MCI subjects.

In a similar manner to patients with damage to neocortex or medial temporal lobe (chapter 1), patients with Alzheimer's disease were deficient in learning the stylus maze (Brouwers et al., 1984). An impairment was also reported in mazes presented on a touch-screen monitor (Ott et al., 2008). The maze scores were linearly correlated with a test of driving competence, a result which underlines the construct validity of this test. Likewise, Alzheimer patients were more likely than controls to be disoriented in places requiring whole body movements, including driving along the streets with a motored vehicle (Uc et al., 2004) and walking in a circular arena (Hort et al., 2007) or a hospital ward (Monacelli et al., 2003). The circular arena task akin to the Morris water maze used in research with animals was deficient in MCI subjects (Hort et al., 2007).

REFERENCES

Alescio-Lautier, B; Michel, BF; Herrera, C; Elahmadi, A; Chambon, C; Touzet, C; Paban, V. Visual and visuospatial short-term memory in mild cognitive impairment and Alzheimer disease: role of attention. *Neuropsychologia,* 2007, 45, 1948-60.

Amieva, H; Jacqmin-Gadda, H; Orgogozo, JM; Le Carret, N; Helmer, C; Letenneur, L; Barberger-Gateau, P; Fabrigoule, C; Dartigues, JF. The 9 year cognitive decline before dementia of the Alzheimer type: a prospective population-based study. *Brain,* 2005, 128, 1093-11.

Anderson, AJ; Stoltzner, S; Lai, F; Su, J; Nixon, RA. Morphological and biochemical assessment of DNA damage and apoptosis in Down syndrome and Alzheimer's disease, and effect of postmortem tissue archival on TUNEL. *Neurobiol. Aging,* 2000, 21, 511-24.

Archer, HA; Edison, P; Brooks, DJ; Barnes, J; Frost, C; Yeatman, T; Fox, NC; Rossor, MN. Amyloid load and cerebral atrophy in Alzheimer's disease: An ^{11}C-PIB positron emission tomography study. *Ann. Neurol., 2006,* 60, 145-7.

Arriagada, PV; Growdon, JH; Hedley-Whyte, T; Hyman, BT. Neurofibrillary tangles but not senile plaques parallel duration and severity of Alzheimer's disease. *Neurology,* 1992, 42, 631-9.

Atack, JR; Perry, EK; Bonham, JR; Perry, RH; Tomlinson, BE; Blessed, G; Fairbairn, A. Molecular forms of acetylcholinestrase in senile dementia of Alzheimer type: selective loss of the intermediate (10S) form. *Neurosci. Lett.,* 1983, 40, 199-204.

Baddeley, AD; Bressi, S; Della Scala, S; Spinnler, H. The decline of working memory in Alzheimer's disease. *Brain,* 1991, 114, 2521-42.

Bartzokis, G; Cummings, JL; Sultzer, D; Henderson, VW; Nuechterlein, KH; Mintz, J. White matter structural integrity in healthy aging adults and patients with Alzheimer disease: a magnetic resonance imaging study. *Arch. Neurol.,* 2003, 60, 393-8.

Baskin, DS; Browning, JL; Pirozzolo, FJ; Korporaal, S; Baskin, JA; Appel, SH. Brain choline acetyltransferase and mental function in Alzheimer disease. *Arch. Neurol.,* 1999, 56, 1121-3.

Becker, JT; Huff, FJ; Nebes, RD; Holland, A; Boller, F. Neuropsychological function in Alzheimer's disease: pattern of impairment and rates of progression. *Arch. Neurol.,* 1988, 45, 263-8.

Bierer, LM; Hof, PR; Purohit, DP; Carlin, L; Schmeidler, J; Davis, KL; Perl, DP. Neocortical neurofibrillary tangles correlate with dementia severity in Alzheimer's disease. *Arch. Neurol.,* 1995, 52, 81-8.

Bird, CM; Chan, D; Hartley, T; Pijnenburg, YA; Rossor, MN; Burgess, N. Topographical short-term memory differentiates Alzheimer's disease from frontotemporal lobar degeneration. *Hippocampus,* 2010, 20, 1154-69.

Blusztajn, JK; Berse, B. The cholinergic neuronal phenotype in Alzheimer's disease. *Met. Brain Dis.,* 2000, 15, 45-64.

Bohnen, NI; Kaufer, DI; Hendrickson, R; Ivanco, LS; Lopresti, B; Davis, JG; Constantine, G; Mathis, CA; Moore, RY; DeKosky, ST. Cognitive correlates of alterations in acetylcholinesterase in Alzheimer's disease. *Neurosci. Lett.,* 2005, 380, 127-32.

Bracco, L; Amaducci, L; Pedone, D; Bino, G; Lazzaro, MP; Carella, F; D'Antona, R; Gallato, R; Denes, G. Italian multicentre study on dementia (SMID): a neuropschological test battery for assessing Alzheimer's disease. *J. Psychiatry Res.,* 1990, 24, 213-26.

Broe, M; Shephard, CE; Milward, EA; Halliday, GM. Relationship between DNA fragmentation, morphological changes and neuronal loss in Alzheimer's disease and dementia with Lewy bodies. *Acta. Neuropathol.,* 2001, 101, 616-24.

Brouwers, P; Cox, C; Martin, A; Chase, T; Fedio, P. Differential perceptual-spatial impairment in Huntington's and Alzheimer's disease. *Arch. Neurol.,* 1984, 41, 1073-6.

Callen, DJA; Black, SE; Gao, F; Caldwell, CB; Szalai, JP. Beyond the hippocampus: MRI volumetry confirms widespread limbic atrophy in AD. *Neurology,* 2001, 57, 1669-74.

Campion, D; Brice, A; Dumanchin, C; Puel, M; Baulac, M; De la Sayette, V; Hannequin, D; Duyckaerts, C; Michon, A; Martin, C; Moreau, V; Penet, C; Martinez, M; Clerget-Darpoux, F; Agid, Y; Frébourg, T. A novel *presenilin 1* mutation resulting in familial Alzheimer's disease with an onset age of 29 years. *NeuroReport,* 1996, 7, 1582-4.

Campion, D; Dumanchin, C; Hannequin, D; Dubois, B; Bellliard, S; Puel, M; Thomas-Anerion, C; Michon, A; Martin, C; Charbonnier, F; Raux, G; Camuzat, A; Penet, C; Mesnage, V; Martinez, M; Clerget-Darpoux, F; Frébourg, T. Early-onset autosomal dominant Alzheimer disease: prevalence, genetic heterogeneity, and mutation spectrum. *Am. J. Hum. Genet.,* 1999, 65, 664-70.

Cherrier, MM; Tröster, A. Geographical remote memory and verbal fluency in Alzheimer's disease. *Dev. Neuropsychol.,* 1998, 14, 631-41.

Collette, F; Van der Linden, M; Salmon, E. Executive dysfunction in Alzheimer's disease. *Cortex,* 1999, 35, 57-72.

Colurso, GJ; Nilson, JE; Vervoort, LG. Quantitative assessment of DNA fragmentation and beta-amyloid deposition in insular cortex and midfrontal gyrus from patients with Alzheimer's disease. *Life Sci.,* 2003, 73, 1795-803.

Coyle, JT; Price, DL; DeLong, MR. Alzheimer's disease: a disorder of cortical cholinergic innervation. *Science,* 1983, 219, 1184-90.

De Jong, LW; van der Hiele, K; Veer, IM; Houwing, JJ; Westendorp, RGJ; Bollen, ELEM; de Bruin, PW; Middelkoop, HAM; van Buchem, MA; van der Grond, J. Strongly reduced volumes of putamen and thalamus in Alzheimer's disease: an MRI study. *Brain,* 2008, 131, 3277-85.

De Jonghe, C; Van Broeckhoven, C. Techniques in the behavioral and neural sciences. In: W Crusio, R Gerlai (eds) *Handbook of molecular-genetic techniques for brain and behavior research,* vol 13. Amsterdam: Elsevier, pp 841-62, 1999.

DeKosky, ST; Harbaugh, RE; Schmitt, FA; Bakay, RAE; Chui, HC; Knopman, DS; Reeder, TM; Shetter, AG; Senter, HJ; Markesbery, WR. Cortical biopsy in Alzheimer's disease: diagnostic accuracy and neurochemical, neuropathological, and cognitive correlations. *Ann. Neurol.,* 1992, 32, 625-32.

Dinamarca, MC; Arrázola, M; Toledo, E; Cerpa, WF; Hancke, J; Inestrosa, NC. Release of acetylcholinesterase (AChE) from β-amyloid plaques assemblies improves the spatial memory impairments in APP-transgenic mice. *Chem. Biol. Interact.,* 2008, 175, 142-49.

Double, KL; Halliday, GM; Kril, JJ; Harasty, JA; Cullen, K; Brooks, WS; Creasey, H; Broe, GA. Topography of brain atrophy during normal aging and Alzheimer's disease. *Neurobiol. Aging,* 1996, 17, 513-21.

Duara, R; Loewenstein, DA; Shen, Q; Barker, W; Greig, MT; Varon, D; Murray, ME; Dickson, DW. Regional patterns of atrophy on MRI in Alzheimer's disease: neuropsychological features and progression rates in the ADNI cohort. *Adv. Alz. Dis.,* 2013, 2, 135-47.

Eikelenboom, P; Bate, C; Van Gool, WA; Hoozemans, JJM; Rozemuller, JM; Veerhuis, R; Williams, A. Neuroinflammation in Alzheimer's disease and prion disease. *Glia,* 2002, 40, 232-9.

Efange, SM; Garland, EM; Staley, JK; Khare, AB; Mash, DC. Vesicular acetylcholine transporter density and Alzheimer's disease. *Neurobiol. Aging,* 1997, 18, 407-13.

Eslinger, PJ; Damasio, AR. Preserved motor learning in Alzheimer's disease: implications for anatomy and behavior. *J. Neurosci.,* 1986, 6, 3006-9.

Flicker, C; Ferris, SH; Reisberg, B. Mild cognitive impairment in the elderly: predictors of dementia. *Neurology,* 1991, 41, 1006-9.

Flicker, C; Ferris, SH; Crook, T; Reisberg, B; Bartus, RT. Equivalent spatial-rotation deficits in normal aging and Alzheimer's disease. *J. Clin. Exp. Neuropsychol.,* 1988, 10, 387-99.

Fox, NC; Crum, WR; Scahill, RI; Stevens, JM; Janssen, JC; Rossor, MN. Imaging of onset and progression of Alzheimer's disease with voxel-compression mapping of serial magnetic resonance images. *Lancet,* 2001, 358, 201-5.

Freedman, M; Oscar-Berman, M. Spatial and visual learning deficits in Alzheimer's and Parkinson's disease. *Brain Cogn.,* 1989a, 11, 114-26.

Freedman, M; Oscar-Berman, M. Selective delayed response deficits in Parkinson's and Alzheimer's disease. *Arch. Neurol.,* 1989b, 43, 886-90.

Frisoni, GB; Testa, C; Zorzan, A; Sabattoli, F; Beltramello, A; Laakso, MP. Detection of grey matter loss in mild Alzheimer's disease with voxel based morphometry. *J. Neurol. Neurosurg. Psychiatry,* 2002, 73, 657-64.

Giannakopoulos, P; Herrmann, FR; Bussiere, T; Bouras, C; Kovari, E; Perl, DP; Morrison, JH; Gold, G; Hof, PR. Tangle and neuron numbers, but not amyloid load, predict cognitive status in Alzheimer's disease. *Neurology,* 2003, 60, 1495-1500.

Gibson, GE; Haroutunian, V; Zhang, H; Park, LCH; Shi, Q; Lesser, M; Mohs, RC; Sheu, K-F; Blass, JP. Mitochondrial damage in Alzheimer's disease varies with apolipoprotein E genotype. *Ann. Neurol.,* 2000, 48, 297-303.

Giordani, B; Boivin, MJ; Hall, AL; Foster, NL; Lehtinen, SJ; Bluemlein, LA; Berent, S. The utility and generality of Mini-Mental State Examination scores in Alzheimer's disease. *Neurology,* 1990, 40, 1894-6.

Gold, G; Kovari, E; Corte, G; Herrmann, FR; Canuto, A; Bussiere, T; Hof, PR; Bouras, C; Giannakopoulos, P. Clinical validity of Aβ-protein deposition staging in brain aging and Alzheimer disease. *J. Neuropathol. Exp. Neurol.,* 2001, 60, 946-52.

Gouras, GK; Tsai, J; Naslund, J; Vincent, B; Edgar, M; Checler, F; Greenfield, JP; Haroutunian, V; Buxbaum, JD; Xu, H; Greengard, P; Relkin, NR. Intraneuronal Aβ42 accumulation in human brain. *Am. J. Pathol.,* 2000, 156, 15-20.

Graham, NL; Emery, T; Hodges, JR. Distinctive cognitive profiles in Alzheimer's disease and subcortical vascular dementia. *J. Neurol. Neurosurg. Psychiatry,* 2004, 75, 61-71.

Gregory, GC; Macdonald, V; Schofield, PR; Kril, JJ; Halliday, GM. Differences in regional brain atrophy in genetic forms of Alzheimer's disease. *Neurobiol. Aging,* 2006, 27, 387-93.

Guariglia, CC; Nitrini, R. Topographical disorientation in Alzheimer's disease. *Arq. Neuropsiquiatr.,* 2009, 67, 967-72.

Hardy, J; Selkoe, DJ. The amyloid hypothesis of Alzheimer's disease: progress and problems on the road to therapeutics. *Science,* 2002, 297, 353-6.

Hardy, J; Adolfsson, R; Alafuzoff, I; Bucht, G; Marcusson, J; Nyberg, P; Perdahl, E; Wester, P; Winblad, B. Transmitter deficits in Alzheimer's disease. *Neurochem. Int.,* 1985, 7, 545-63.

Helmes, E; Ostbye, T. Beyond memory impairment: cognitive changes in Alzheimer's disease. *Arch. Clin. Neuropsychol.,* 2002, 17, 179-93.

Henderson, VW; Mack, W; Williams, BW. Spatial disorientation in Alzheimer's disease. *Arch. Neurol.,* 1989, 46, 391-4.

Herholz, K; Ebmeier, K. Clinical amyloid imaging in Alzheimer's disease. *Lancet Neurol.,* 2011, 10, 667-70.

Herholz, K; Weisenbach, S; Zundorf, G; Lenz, O; Schroder, H; Bauer, B; Kalbe, E; Heiss, WD. *In vivo* study of acetylcholine esterase in basal forebrain, amygdala, and cortex in mild to moderate Alzheimer disease. *NeuroImage,* 2004, 21, 136-43.

Hort, J; Laczo, J; Vyhnalek, M; Bojar, M; Bures, J; Vlcek, K. Spatial navigation deficit in amnestic mild cognitive impairment. *Proc. Natl. Acad. Sci. USA,* 2007, 104, 4042-7.

Huntley, J; Bor, D; Hampshire, A; Owen, A; Howard, R. Working memory task performance and chunking in early Alzheimer's disease. *Br. J. Psychiatry,* 2011, 198, 398-403.

Hyman, BT; Kromer, LJ; Van Hoesen, GW. Reinnervation of the hippocampal perforant pathway zone in Alzheimer's disease. *Ann. Neurol.,* 1987, 21, 259-67.

Iachini, I; Iavarone, A; Senese, VP; Ruotolo, F; Ruggiero, G. Visuospatial memory in healthy elderly, AD and MCI: a review. *Curr. Aging Sci.,* 2009, 2, 43-59.

Ingelsson, M; Fukumoto, H; Newell, KL; Growdon, JH; Hedley-Whyte, ET; Frosch, MP; Albert, MS; Hyman, BT; Irizarry, MC. Early Aβ accumulation and progressive synaptic loss, gliosis, and tangle formation in AD brain. *Neurology,* 2004, 62, 925-31.

Iwatsubo, T; Odaka, A; Suzuki, N; Mizusawa, H; Nukina, N; Ihara, Y. Visualization of Aβ42(43) and Aβ40 in senile plaques and end-specific Aβ monoclonals: evidence that an initially deposited species is Aβ42(43). *Neuron,* 1994, 13, 45-53.

Jack, CR; Petersen, RC; Xu, Y; O'Brien, PC; Smith, GE; Ivnik, RJ; Tangalos, EG; Kokmen, E. Rate of medial temporal lobe atrophy in typical aging and Alzheimer's disease. *Neurology,* 1998, 51, 993-99.

Jacobs, DH; Adair, JC; Williamson, DJ; Na, DL; Gold, M; Foundas, AL; Shuren, JE; Cibula, JE; Heilman, KM. Apraxia and motor-skill acquisition in Alzheimer's disease are dissociable. *Neuropsychologia,* 1999, 37, 875-80.

Jacobs, DM; Sano, M; Dooneief, G; Marder, K; Bell, KL; Stern, Y. Neuropsychological detection and characterization of preclinical Alzheimer's disease. *Neurology,* 1995, 45, 957-62.

Jeste, DV; Wragg, RE; Salmon DP; Harris, MJ; Thal, LJ. Cognitive deficits of patients with Alzheimer's disease with and without delusions. *Am. J. Psychiatry,* 1992, 149, 184-9.

Juottonen, K; Laakso, MP; Inausti, R; Lehtovirta, M; Pitänen, A; Partanen, K; Soininen, H. Volumes of the entorhinal and perirhinal cortices in Alzheimer's disease. *Neurobiol. Aging,* 1998, 19, 15-22.

Kalaria, RN; Kroon, SN; Grahovac, I; Perry, G. Acetylcholinesterase and its association with heparan sulphate proteoglycans in cortical amyloid deposits of Alzheimer's disease. *Neuroscience,* 1992, 51, 177-84.

Kanne, SM; Balota, DA; Storandt, M; McKeel, DW Jr; Morris, JC. Relating anatomy to function in Alzheimer's disease: neuropsychological profiles predict regional neuropathology 5 years later. *Neurology,* 1998, 50, 979-85.

Karas, GB; Scheltens, P; Rombouts, SARB; Visser, PJ; van Schijndel, RA; Fox, NC; Markhof, F. Global and local gray matter loss in mild cognitive impairment and Alzheimer's disease. *NeuroImage,* 2004, 23, 708-16.

Karlsson, T; Börjesson, A; Adolfsson, R; Nilsson, L-G. Successive memory test performance and priming in Alzheimer's disease: evidence from the word-fragment completion task. *Cortex,* 2002, 38, 341-55.

Kavcic, V; Fernandez, R; Logan, D; Duffy, C. Neurophysiological and perceptual correlates of navigational impairment in Alzheimer's disease. *Brain,* 2006, 129, 736-46.

Kessels, RP; Feijen, J; Postma, A. Implicit and explicit memory for spatial information in Alzheimer's disease. *Dement. Geriatr. Cogn. Disord.,* 2005, 20, 184-91.

Kokmen, E; Smith, GE; Petersen, RC; Tangalos, E; Ivnik, RC. The Short Test of Mental Status: correlations with standardized psychometric testing. *Arch. Neurol.,* 1991, 48, 725-8.

Kuhl, DE; Koeppe, RA; Minoshima, S; Snyder, SE; Ficaro, EP; Foster, NL; Frey, KA; Kilbourn, MR; Frey, KA; Kilbourn, MR. *In vivo* mapping of cerebral acetylcholinesterase activity in aging and Alzheimer's disease. *Neurology,* 1999, 52, 691-9.

Kumar-Singh, S; Theuns, J; Van Broeck, B; Pirici, D; Venneken, K; Corsmit, E; Cruts, M; Dermaut, B; Wang, R; Van Broeckhoven, C. Mean age-of-onset of familial alzheimer disease caused by presenilin mutations correlates with both increased Aβ42 and decreased Aβ40. *Hum. Mutat.,* 2006, 27, 686-95.

Laakso, MP; Soininen, H; Partanen, K; Lehtovirta, M; Hallikainen, M; Hänninen, T; Helkala, E-L; Riekkinen, PJ Jr. MRI of the hippocampus in Alzheimer's disease: sensitivity, specificity, and analysis of the incorrectly classified subjects. *Neurobiol. Aging,* 1998, 19, 23-31.

Larner, AJ; Doran, M. Clinical phenotypic heterogeneity of Alzheimer's disease associated with mutations of the presenilin-1 gene. *J. Neurol.,* 2006, 253, 139-58.

Lassmann, H; Bancher, C; Breitschopf, H; Wegiel, J; Bobinki, M; Jellinger, K. Cell death in Alzheimer's disease evaluated by DNA fragmentation *in situ. Acta Neuropathol.,* 1995, 89, 35-41.

Lee, AC; Rahman, S; Hodges, JR; Sahakian, BJ, Graham, KS. Associative and recognition memory for novel objects in dementia: implications for diagnosis. *Eur. J. Neurosci.,* 2003, 18, 1660-70.

Lyness, SA; Zarow, C; Chui, HC. Neuron loss in key cholinergic and aminergic nuclei in Alzheimer's disease: a meta-analysis. *Neurobiol. Aging,* 2003, 24, 1-23.

Maki, Y; Yoshida, H; Yamaguchi, H. Computerized visuo-spatial memory test as a supplementary screening test for dementia. *Psychogeriatrics,* 2010, 10, 77-82.

Mapstone, M; Steffenella, TM; Duffy, CJ. A visuospatial variant of mild cognitive impairment: getting lost between aging and AD. *Neurology,* 2003, 60, 802-8.

Marra, C; Ferraccioli, M; Vita, MG; Quaranta, D; Gainotti, G. Patterns of cognitive decline and rates of conversion to dementia in patients with degenerative and vascular forms of MCI. *Curr. Alzheimer. Res.,* 2011, 8, 24-31.

Martin, EM; Wilson, RS; Penn, RD; Fox, JH; Clasen, RA; Savoy, SM. Cortical biopsy results in Alzheimer's disease: correlation with cognitive deficits. *Neurology,* 1987, 37, 1201-4.

McKhann, G; Drachman, D; Folstein, M; Katzman, R; Price, D; Stadlan, EM. Clinical diagnosis of Alzheimer's disease: report of the NINCDS-ADRA work group under the Department of Health and Human Services task force on Alzheimer's disease. *Neurology,* 1984, 34, 939-44.

McLean, CA; Cherny, RA; Fraser, FW; Fuller, SJ; Smith, MJ; Beyreuther, K; Bush, AI; Masters, CL. Soluble pool of Aβ amyloid as a determinant of severity of neurodegeneration in Alzheimer's disease. *Ann. Neurol.,* 1999, 46, 860-6.

Mendez, MF; Underwood, KL; Zander, BA; Mastri, AR; Sung, JH; Frey, WH II. Risk factors in Alzheimer's disease: a clinicopathologic study. *Neurology,* 1992, 42, 770-5.

Merched, A; Xia, Y; Visvikis, S; Serot, JM; Siest, G. Decreased high-density lipoprotein cholesterol and serum apolipoprotein AI concentrations are highly correlated with the severity of Alzheimer's disease. *Neurobiol. Aging,* 2000, 21, 27-30.

Minger, SL; Honer, WG; Esiri, MM; McDonald, B; Keene, J; Nicoll, JA; Carter, J; Hope, T; Francis, PT. Synaptic pathology in prefrontal cortex is present only with severe dementia in Alzheimer disease. *J. Neuropathol. Exp. Neurol.,* 2001, 60, 929-36.

Mitew, S; Kirkcaldie, MT; Halliday, GM; Shepherd, CE; Vickers, JC; Dickson, TC. Focal demyelination in Alzheimer's disease and transgenic mouse models. *Acta Neuropathol.,* 2010, 119, 567-77.

Mohr, E; Litvan, I; Williams, J; Fedio, Chase, T. Selective deficits in Alzheimer and Parkinsonian dementia: visuospatial function. *Can. J. Neurol. Sci.,* 1990, 17, 292-7.

Monacelli, AM; Cushman, LA; Kavcic, V; Duffy, CJ. Spatial disorientation in Alzheimer's disease: the remembrance of things passed. *Neurology,* 2003, 61, 1491-7.

Moran, MA; Mufson, EJ; Gomez-Ramos, P. Colocalization of cholinesterases with β amyloid protein in aged and Alzheimer's brain. *Acta Neuropathol.,* 1993, 85, 362-9.

Morris, JC. Relationship between plaques and tangles to Alzheimer's disease phenotype. In: A Goate, F Ashall (eds) *Pathobiology of Alzheimer's disease.* San Diego: Academic Press, pp. 193-223, 1995.

Morris, JC; Heyman, A; Mohs, RC; Hughes, JP; van Belle, G; Fillenbaum, G; Mellits, ED; Clark, C. The Consortium to Establish a Registry for Alzheimer's Disease (CERAD). Part I. Clinical and neuropsychological assessment of Alzheimer's disease. *Neurology,* 1989, 39, 1159-65.

Mosconi, L; Sorbi, S; Nacmias, B; De Cristofaro, MT; Fayyaz, M; Cellini, E; Bagnoli, S; Bracco, L; Herholz, K; Pupi, A. Brain metabolic differences between sporadic and familial Alzheimer's disease. *Neurology,* 2003, 61, 1138-40.

Näslund, J; Haroutunian, V; Mohs, R; Davis, KL; Davies, P; Greengard, P; Buxbaum, JD. Correlation between elevated levels of amyloid β-peptide in the brain and cognitive decline. *JAMA,* 2000, 283, 1571-7.

Neve, RL; McPhie, DL; Chen, Y. Alzheimer's disease: a dysfunction of the amyloid precursor protein. *Brain Res.,* 2000, 886, 54-66.

Nielson, KA; Cummings, BJ; Cotman, C.W. Constructional apraxia in Alzheimer's disease correlates with neuritic neuropathology in occipital cortex. *Brain Res.,* 1996, 741, 284-93.

Nissen, MJ; Corkin, S; Buonanno, FS; Growden, JH; Wray, SH; Bauer, J. Spatial vision in Alzheimer's disease: general findings and a case report. *Arch. Neurol.,* 1985, 42, 667-71.

Nordahl, CW; Ranganath, C; Yonelinas, AP; DeCarli, C; Reed, BR; Jagust, WJ. Different mechanisms of episodic memory failure in mild cognitive impairment. *Neuropsychologia,* 2005, 43, 1688-97.

Okamura, N; Funaki, Y; Tashiro, M; Kato, M; Ishikawa, Y; Maruyama, M; Ishikawa, H; Meguro, K; Iwata, R; Yanai, K. *In vivo* visualization of donepezil binding in the brain of patients with Alzheimer's disease. *Br. J. Clin. Pharmacol.,* 2008, 65, 472-9.

Ott, BR; Festa, EK; Amick, MM; Grace, J; Davis, JD; Heindel, WC. Computerized maze navigation and on-road performance by drivers with dementia. *J. Geriatr. Psychiatry Neurol.,* 2008, 21, 18-25.

Palmer, AM; Francis, PT; Bowen, DM; Benton, JS; Neary, D; Mann, DMA; Snowden, JS. Catecholaminergic neurones assessed ante-mortem in Alzheimer's disease. *Brain Res.,* 1987, 414, 365-75.

Pappas, BA; Bayley, PJ; Bui, BK; Hansen, LA; Thal, LJ. Choline acetyltransferase activity and cognitive domain scores of Alzheimer's patients. *Neurobiol. Aging,* 2001, 21, 11-7.

Parasuraman, R; Greenwood, PM; Alexander, GE; Parasuraman, R; Greenwood, PM; Alexander, GE. Alzheimer disease constricts the dynamic range of spatial attention in visual search. *Neuropsychologia,* 2000, 38, 1126-35.

Parvathy, S; Davies, P; Haroutunian, V; Purohit, DP; Davis, KL; Mohs, RC; Park, H; Moran, TM; Chan, JY; Buxbaum, JD. Correlations between $A\beta_x$-40, $A\beta_x$-42, and $A\beta_x$-43-containing amyloid plaques and cognitive decline. *Arch. Neurol.,* 2001, 58, 2025-32.

Pengas, G; Patterson, K; Arnold, RJ; Bird, CM; Burgess, N; Nestor, PJ. Lost and found: bespoke memory testing for Alzheimer's disease and semantic dementia. *J. Alzheimers Dis,* 2010, 21, 1347-65.

Perl, DP. Neuropathology of Alzheimer's disease and related disorders. *Neurol. Clin.,* 2000, 18, 847-64.

Perry, G; Cras, P; Siedlak, SL; Tabaton, M; Kawai, M. Beta protein immunoreactivity is found in the majority of neurofibrillary tangles of Alzheimer's disease. *Am. J. Pathol.,* 1992, 140, 283-90.

Pillot, T; Drouet, B; Queillé, S; Labeur, C; Vandekerckhove, J; Rosseneu, M; Pinçon-Raymond, M; Chambaz, J. The nonfibrillar β-amyloid peptide induces apopoptic neuronal cell death: involvement of its C-terminal fusogenic domain. *J. Neurochem.,* 1999, 73, 1626-34.

Possin, KL; Laluz, VR; Alcantar, OZ; Miller, BL; Kramer, JH. Distinct neuroanatomical substrates and cognitive mechanisms of figure copy performance in Alzheimer's disease and behavioral variant frontotemporal dementia. *Neuropsychologia,* 2011, 49, 43-8.

Raina, AK; Hochman, A; Zhu, X; Rottkamp, A; Nunomura, A; Siedlak, SL; Boux, H; Castellani, R; Perry, G; Smith, MA. Abortive apoptosis in Alzheimer's disease. *Acta Neuropathol.,* 2001, 101, 305-10.

Rainville, C; Marchand, N; Passini, R. Performances of patients with a dementia of the Alzheimer type in the Standardized Road-Map test of Direction Sense. *Neuropsychologia,* 2002a, 40, 567-73.

Rainville, C; Amieva, H; Lafont, S; Dartigues, JF; Orgogozo, JM; Fabrigoule, C. Executive function deficits in patients with dementia of the Alzheimer's type: a study with a Tower of London task. *Arch. Clin. Neuropsychol.,* 2002b, 17, 513-30.

Reddy, PH; Mani, G; Park, BS; Jacques, J; Murdoch, G; Whetsell, W Jr; Kaye, J; Manczak, M. Differential loss of synaptic proteins in Alzheimer's disease: implications for synaptic dysfunction. *J. Alzheimers Dis.,* 2005, 7, 103-17.

Riemenschneider, M; Schmolke, M; Lautenschlager, N; Vanderstichele, H; Vanmechelen, E; Guder, WG; Kurz, A. Association of CSF apolipoprotein E, Aβ42 and cognition. *Neurobiol. Aging,* 2002, 23, 205-11.

Rinne, JO; Kaasinen, V; Jarvenpää, T; Nagren, K; Roivainen, A; Yu, M; Oikonen, V; Kurki, T. Brain acetylcholinesterase activity in mild cognitive impairment and early Alzheimer's disease. *J. Neurol. Neurosurg. Psychiatry,* 2003, 74, 113-5.

Rogers, JD; Brogan, D; Mirra, S. The nucleus basalis of Meynert in neurological disease: a quantitative morphological study. *Ann. Neurol.,* 1985, 17, 163-70.

Rohan de Silva, HA; Jen, A; Wickenden, C; Jen, L-S; Wilkinson, SL; Patel, AJ. Cell-specific expression of small β-amyloid precursor protein isoform mRNAs and proteins in neurons and astrocytes. *Mol. Brain Res.,* 1997, 47, 147-56.

Rombouts, SARB; Barkhof, F; Witter, MP; Scheltens, P. Unbiased whole-brain analysis of gray matter loss in Alzheimer's disease. *Neurosci. Lett.*, 2000, 285, 231-3.

Rovelet-Lecrux, A; Hannequin, D; Raux, G; Le Meur, N; Laquerriere, A; Vital, A; Dumanchin, C; Feuillette, S; Brice, A; Vercelletto, M; Dubas, F; Frébourg, T; Campion, D. APP locus duplication causes autosomal dominant early-onset Alzheimer disease with cerebral amyloid angiopathy. *Nature Genet.,* 2006, 38, 24-6.

Samuel, W; Terry, RD; DeTeresa, R; Butters, N; Masliah, E. Clinical correlates of cortical and nucleus basalis pathology in Alzheimer dementia. *Arch. Neurol.,* 1994, 51, 772-8.

Scahill, RI; Schott, JM; Stevens, JM; Rossor, MN; Fox, NC. Mapping the evolution of regional atrophy in Alzheimer's disease: unbiased analysis of fluid-registered serial MRI. *Proc. Natl. Acad. Sci. USA*, 2002, 99, 4703-7.

Serra, L; Fadda, L; Perri, R; Spanò, B; Marra, C; Castelli, D; Torso, M; Makovac, E; Cercignani, M; Caltagirone, C; Bozzali, M. Constructional apraxia as a distinctive cognitive and structural brain feature of pre-senile Alzheimer's disease. *J. Alzheimers Dis.,* 2014, 38, 391-402.

Shinotoh, H; Namba, H; Fukushi, K; Nagatsuka, S; Tanaka, N; Aotsuka, A; Ota, T; Tanada, S; Irie, T. Progressive loss of cortical acetylcholinesterase activity in association with cognitive decline in Alzheimer's disease: a positron emission tomography study. *Ann. Neurol.,* 2000, 48, 194-200.

Sjöbeck, M; Englund, E. Alzheimer's disease and the cerebellum: a morphologic study on neuronal and glial changes. *Dement. Geriatr. Cogn. Disord.,* 2001, 12, 211-8.

Soreq, H; Seidman, S. Acetylcholinesterase: new roles for an old actor. *Nature Rev. Neurosci.,* 2001, 2, 294-302.

Spinnler, H; Della Scala, S; Bandera, R; Baddeley, A. Dementia, ageing, and the structure of human memory. *Cogn. Neuropsychology,* 1988, 5, 193-211.

Su, JH; Kesslak, JP; Head, E; Cotman, CW. Caspase-cleaved amyloid precursor protein and activated caspase-3 are co-localized in the granules of granulovacuolar degeneration in Alzheimer's disease and Down's syndrome brain. *Acta Neuropathol.,* 2002, 104, 1-6.

Swearer, JM; O'Donnell, BF; Drachman, DA; Woodward, BM. 1992. Neuropsychological features of familial Alzheimer's disease. *Ann. Neurol.,* 1992, 32, 687-94.

Takahashi, RH; Milner, TA; Li, F; Nam, EE; Edgar, MA; Yamaguchi, H; Beal, MF; Xu, H; Greengard, P; Gouras, GK. Intraneuronal Alzheimer Aβ42 accumulates in multivesicular bodies and is associated with synaptic pathology. *Am. J. Pathol.,* 2002, 161, 1869-79.

Tales, A; Muir, J; Jones, R; Bayer, A; Snowden, RJ. The effects of saliency and task difficulty on visual search performance in ageing and Alzheimer's disease. *Neuropsychologia,* 2004, 42, 335-45.

Talesa, VN. Acetylcholinesterase in Alzheimer's disease. *Mech. Aging Dev.,* 2001, 122, 1961-9.

Tanzi, RE; Bertram, L. New frontiers in Alzheimer's disease genetics. *Neuron,* 2001, 32, 181-4.

Tekin, S; Mega, MS; Masterman, DM; Chow, T; Garakian, J; Vinters, HV; Cummings, JL. Orbitofrontal and anterior cingulate cortex neurofibrillary tangle burden is associated with agitation in Alzheimer disease. *Ann. Neurol.,* 2001, 49, 355-61.

Terry, RD; Masliah, E; Salmon, DP; Butters, N; DeTeresa, R; Hill, R; Hansen, LA; Katzman, R. Physical basis of cognitive alterations in Alzheimer's disease: synapse loss is the major correlate of cognitive impairment. *Ann. Neurol.,* 1991, 30, 572-80.

Tetewsky, SJ; Duffy, CJ. Visual loss and getting lost in Alzheimer's disease. *Neurology,* 1999, 52, 958-65.

Thal, DR; Rüb, U; Schultz, C; Sassin, I; Ghebremedhin, E; Tredici, KD; Braak, E; Braak, H. Sequence of Aβ-protein deposition in the human medial temporal lobe. *J. Neuropathol. Exp. Neurol.,* 2000, 59, 733-48.

Tiraboschi, P; Hansen, L.A; Thal, LJ; Corey-Bloom, J. The importance of neuritic plaques and tangles to the development and evolution of AD. *Neurology,* 2004, 62, 1984-9.

Toepper, M; Beblo, T; Thomas, C; Driessen, M. Early detection of Alzheimer's disease: a new working memory paradigm. *Int. J. Geriatr. Psychiatry,* 2008, 23, 272-8.

Troncoso, JC; Sukhov, RR; Kawas, CH; Koliatsos, VE. *In situ* labeling of dying cortical neurons in normal aging and in Alzheimer's disease: correlations with senile plaques and disease progression. *J. Neuropathol. Exp. Neurol.,* 1996, 55, 1134-42.

Turner, PR; O'Connor, K; Tate, WP; Abraham, WC. Roles of amyloid precursor protein and its fragments in regulating neural activity, plasticity and memory. *Prog. Neurobiol.,* 2003, 70, 1-32.

Tyler, SJ; Dawbarn, D; Wilcock, GK; Allen, SJ. α- and β-secretase: profound changes in Alzheimer's disease. *Biochem. Biophys. Res. Comm.,* 2002, 299, 373-6.

Uc, EY; Rizzo, M; Anderson, SW; Shi, Q; Dawson, JD. Driver route-following and safety errors in early Alzheimer disease. *Neurology,* 2004, 63, 832-7.

Ueki, A; Goto, K; Sato, N; Iso, H; Morita, Y. Prepulse inhibition of acoustic startle response in mild cognitive impairment and mild dementia of Alzheimer type. *Psychiatry Clin. Neurosci.,* 2006, 60, 55-62.

Van Beek, AH; Claassen, JA. The cerebrovascular role of the cholinergic neural system in Alzheimer's disease. *Behav. Brain Res.,* 2011, 221, 537-42.

von Bernhardi, R; Ramírez, G; De Ferrari, GV; Inestrosa, NC. Acetylcholinesterase induces the expression of the small β-amyloid precursor protein in glia and activates glial cells in culture. *Neurobiol. Dis.,* 2003, 14, 447-57.

Weintraub, S; Mesulam, MM. Four neuropsychological profiles in dementia. In: H Spinnler, F Boller (eds) *Handbook of neuropsychology,* vol 8. Amsterdam: Elsevier, pp 253-82, 1993.

Whitford, GW. Alzheimer's disease and serotonin: a review. *Neuropsychobiology,* 1986, 15, 133-42.

Wiechmann, A; Hall, JR; O'Bryant, SE. The utility of the spatial span in a clinical geriatric population. *Neuropsychol. Dev. Cogn. B. Aging Neuropsychol. Cogn.,* 2011, 18, 56-63.

Wolfe, MS; Xia, W; Ostaszewki, BL; Diehl, TS; Kimberly, WT; Selkoe, DJ. Two transmembrane aspartates in presenilin-1 required for presenilin endoproteolysis and γ-secretase activity. *Nature,* 1999, 398, 513-7.

Woodruff-Pak, DS; Papka, M. Alzheimer's disease and eyeblink conditioning: 750 ms trace vs. 400 ms delay paradigm. *Neurobiol. Aging,* 1996, 17, 397-404.

Woodruff-Pak, DS; Romano, S; Papka, M. Training to criterion in eyeblink classical conditioning in Alzheimer's disease, Down's syndrome with Alzheimer's disease, and healthy elderly. *Behav. Neurosci.,* 1996, 110, 22-9.

Wu, C-K; Thal, L; Pizzo, D; Hansen, L; Masliah, E; Geula, C. Apoptotic signals within the basal forebrain cholinergic neurons in Alzheimer's disease. *Exp. Neurol.,* 2005, 195, 484-96.

Yamaguchi, H; Hirai, S; Morimatsu, M; Shoji, M; Ihara, Y. A variety of cerebral amyloid deposits in the brains of the Alzheimer-type dementia demonstrated by beta protein immunostaining. *Acta Neuropathol.,* 1988, 76, 541-9.

Yamaguchi, H; Maat-Schieman, MLC; van Duinen, SG; Prins, FA; Neeskens, P; Natté, R; Roos, RAC. Amyloid b protein (Ab) starts to deposit as plasma membrane-bound form in diffuse plaques of brains from hereditary cerebral hemorrhage with amyloid-Dutch type, Alzheimer disease and nondemented aged subjects. *J. Neuropathol. Exp. Neurol.,* 2000, 59, 723-32.

In: Spatial Orientation in Humans and Animals
Editor: Robert Lalonde
ISBN: 978-1-63463-001-6
© 2014 Nova Science Publishers, Inc.

Chapter 5

SPATIAL DEFICITS ON *APP* TRANSGENIC MOUSE MODELS OF ALZHEIMER'S DISEASE

***Robert Lalonde*[1*]*, K. Fukuchi*[2†] *and Catherine Strazielle*[3‡]**
[1]Université de Rouen, Dépt. Psychologie, Mont-Saint-Aignan Cedex France
[2]University of Illinois College of Medicine at Peoria,
Department of Cancer Biology and Pharmacology, Peoria IL
[3]Université de Lorraine, Laboratoire "Stress, Immunité, Pathogènes"
and Service de Microscopie Electronique, Faculté de Médecine, and CHU de Nancy,
Vandoeuvre-les-Nancy, France

1. PATHOLOGICAL CHARACTERISTICS OF *APP* MUTANTS

Thanks to the discovery of pathologic mutations in humans (chapter 4), several Alzheimer hallmarks have been attained in the engineering of transgenic mice (Cole and Frautschy, 1997; Dodart et al., 2002; Duff, 1999; German and Eisch, 2004; Hock and Lamb, 2001; Hsiao, 1998a,b; Janus and Westaway, 2001; Seabrook and Rosahl 1999; van Leuven, 2000). The brains of *APP* models most often resemble rare cases of Alzheimer dementia with high levels of Aβ protein and hyperphosphorylated tau but without neurofibrillary tangles (Tiraboschi et al., 2004a,b). The genetic characteristics of mice expressing mutated or wild-type *APP* or its C-terminal fragments are presented in Table 1.

2. MORRIS WATER MAZE

In the Morris water maze (chapter 2), mice placed in different areas of a water basin reach a platform found at a constant location, either submerged (place learning), or visible (visuomotor control). Although several reviews have appeared concerning the behavior of

* Email: robert.lalonde@univ-rouen.fr
† Email: kfukuchi@uic.edu
‡ Email: catherine.strazielle@univ-lorraine.fr

Alzheimer-like models on the Morris water maze (Bromley-Brits et al., 2011; Karl et al., 2012; McDonald and Overmier, 1998; Morgan, 2003; Morris et al., 1982; Phinney et al., 2003; Scearce-Levie, 2011; Stewart et al., 2011; Tanila, 2012), the present one is the most thorough and comprehensive (Table 2).

Table 1. Description of behaviorally characterized transgenic mice expressing human mutated *APP genes* or wild-type (WT) with or without mutated *PS1* or Aβ$_{42}$

Name	*APP* Mutation	Promoter	Brain Aβ plaques	References
APP$_{695}$SWE Tg2576	transgenic Swedish	hamster *Prp*	yes	Hsiao et al., 1996
APP$_{695}$SWEch	transgenic Swedish	murine *Prp*	no	Borchelt et al., 1996, 1997
APP$_{695}$SWE/*Thy1* TgTAS10	transgenic Swedish	murine *Thy1*	yes	Richardson et al., 2003
APP$_{751}$SWE TgAPP23	transgenic Swedish	murine *Thy1*	yes	Sturchler-Pierrat et al., 1997
APP/IND TgPDAPP	transgenic Indiana	platelet-derived *PDGFβ*	yes	Games et al., 1995
APP$_{695}$SWE/IND TgCRND8	transgenic Swedish + Indiana	hamster *Prp*	yes	Chishti et al., 2001
APP/SWE/IND J9 and J20 lines	transgenic Swedish + Indiana	platelet-derived *PDGFβ*	yes	Mucke et al., 2000; Palop et al., 2003
APP$_{751}$SWE/IND TgAPP-Sw, V717F/B6	transgenic Swedish + Indiana	platelet-derived *PDGFβ*	no	Lee et al., 2004
APP/ArcSwe	transgenic Swedish + Arctic	murine *Thy1*	yes	Lord et al., 2006
APP$_{695}$LD *APP*$_{751}$LD	transgenic London	murine *Thy1*	yes	Moechars et al., 1999
APP/LD-PDGF	transgenic London	platelet-derived *PDGFβ*	yes	Qin et al., 2000
APP/SweLd	transgenic Swedish + London	murine *Prp*	yes	Kitamoto et al., 2002
APP/OSA	transgenic Osaka	murine *Prp*	no	Tomiyama et al., 2010
APP/DU/*THY1*	transgenic Dutch	human *THY1*	no	Andrä et al., 1996
APP$_{695}$WT/*Mt2*	transgenic wild-type	*Mt2*	no	Yamaguchi et al., 1991
APP$_{695}$WT/*Thy1*	transgenic wild-type	murine *Thy1*	no	Moechars et al., 1999
APP$_{751}$WT	transgenic wild-type	rat *Nse*	no	Quon et al., 1991

Name	*APP* Mutation	Promoter	Brain Aβ plaques	References
APP_{751}WT	transgenic wild-type	rat *Nse*	no	Mucke et al., 1994
APP-WT/*PDGFβ*	transgenic wild-type	*PDGFβ*	no	Simón et al., 2009
APP/C104	transgenic C104 fragment	human *NEFL*	yes	Nalbantoglu et al., 1997
APP/C100-Flag	transgenic C100 fragment	*Dmd*	no	Oster-Granite et al., 1996
APP/C99 Tg13592	transgenic C99 fragment	CMVenhancer/ Chicken *Actb*	no	Fukuchi et al., 1996
APP/C99/IND	transgenic C99 fragment+ Indiana	platelet-derived *PDGFβ*	no	Lee et al., 2006b
APP_{695}SWE + *PS1*/M146L	Bigenic	hamster *Prp* + *PDGFβ2*	yes	Holcomb et al., 1998
APP_{695}SWE + *PS1*/A246E	Bigenic	hamster *Prp* + human *THY1*	yes	Dineley et al., 2002
APP_{695}SWEch + *PS1*/A246E	Bigenic	murine *Prp*	yes	Borchelt et al., 1996
APP_{695}SWE + *PS1*/M146V	Bigenic	murine *Thy1*	yes	Howlett et al., 2004
APP_{695}SWEch + *PS1*/ΔE9	Bigenic	murine *Prp*	yes	Jankowsky et al., 2005
APP_{695}SWE/co + *PS1*/ΔE9	co-injected bigenic	murine *Prp*	yes	Jankowsky et al., 2004
APP_{751}SL + *PS1*/M146L	Bigenic	murine *Thy1*+ human *HMGCR*	yes	Blanchard et al., 2003; Casas et al., 2004
*APP*swe + *PS1*/L166P	co-injected bigenic	murine *Thy1*	yes	Radde et al., 2006
3xTg-AD	Trigenic	murine *Thy1*+ murine control + murine *Thy1*+	yes	Oddo et al., 2003
PLB1 including *PS1*/A246E	knock-in K670N/ M671L/ London	murine *Camk2a*	yes	Platt et al., 2011
5xFAD including *PS1*/M146L *PS1*/L286V	transgenic Swedish, London, Florida	murine *Thy1*	yes	Oakley et al., 2006; Ohno et al., 2006
APP	Null	not applicable	no	Müller et al., 1994

2.1. APP Mice with the Swedish Mutation

Relative to wild-type, the acquisition of the hidden platform version of the Morris water maze was slowed down in 9 month-old APP_{695}SWE mutants (Tg2576) with the K670M/K671L Swedish mutation (Hsiao et al., 1996). This result was reproduced in mice ranging from 15 to 26 months of age (Fernandez-Fernandez et al., 2012; King and Arendash, 2002b; Lassalle et al., 2008; Lee et al., 2006a, 2011; Li et al., 2003; Lindner et al., 2006; Nishida et al., 2006; Quinn et al., 2007; Ribé et al., 2005; Ricobaraza et al., 2009). Results are more variable at a younger age, impairments being noted at 3 (Lindner et al., 2006), 4 (Fernandez-Fernandez et al., 2012), 5 (Arendash et al., 2004; Hongpaisan et al., 2011), 8 (Fukumoto et al., 2010), or 12 (Adriani et al., 2006; Bianchi et al., 2008) months, but not at 6 (Holcomb et al., 1999; Zhuo et al., 2007), 8 (Nishida et al., 2006, Ribes et al., 2008), 9 (Ribé et al., 2005), or 12 (Choi et al., 2012; Massaad et al., 2009), 14 (Zhuo et al., 2007), or from 3 to 14 months (King and Arendash, 2002b), so that the overall effect size appears weak (Reed et al., 2010). Since Aβ plaques first appear at 7 months, the behavioral variability is probably caused by variable age-related changes in neuroanatomic circuits in the course of Aβ plaque formation. In addition, methodologic factors likely intervene, such as changes in the genetic background of the mixed strain which include SJL, the length of behavioral training, and the quality of room cues. Other possible sources of variability include the visual impact of APP accumulation on the retina of these mutants (Dutescu et al., 2009), their increased cataracts relative to controls (Melov et al., 2005), and the retinal degeneration inherent on the SJL background homozygous for the $Pde6^{rd1}$ gene (www.jaxmice.org/strain/000686.html).

Age-dependent results are likewise seen during the probe trial, performances being impaired at 4 (Fernandez-Fernandez et al., 2012), 8 (Ribes et al., 2008), 9 (Hsiao et al., 1996), 12 (Massaad et al., 2009), 16 (Lee et al., 2011; Quinn et al., 2007), 17 (Lassalle et al., 2008; Ricobaraza et al., 2009), and 18 (Fernandez-Fernandez et al., 2012) months, but not in the range from 3 to 26 months (Adriani et al., 2006; King and Arendash, 2002b; Lindner et al., 2006; Nishida et al., 2006). But when the probe is interpolated throughout training, APP_{695}SWE mice were uniformly deficient at 5 (Arendash et al., 2004), 9 (Westerman et al., 2002), 16 (Ribé et al., 2005), and 17 (Lee et al., 2006a) months. Variability in results is also a feature of the visible platform subtask. On one hand, APP_{695}SWE mice were impaired at 3, 5, 9, 14, 16, 19, and 26 months (Arendash et al., 2004; King and Arendash, 2002b; Lindner et al., 2006; Ribé et al., 2005), but not at 6, 9, and 16 months (Holcomb et al., 1999; Li et al., 2003; Lee et al., 2006a; Nishida et al., 2006; Quinn et al., 2007). The length of performance is a likely source of variability, since in two reports (Arendash et al., 2004; Ribé et al., 2005), the mutants were impaired at the initial but not the final trials.

Since the single transgenic shows variability, it is no surprise to note the same pattern in the APP_{695}SWE+*PS1*/M146L bigenic from which it is derived, impairments in acquisition being found at 8 (Sadowski et al., 2004), 15 (Arendash et al., 2001a), and 22 (Sadowski et al., 2004) months of age, but not at 5 (Arendash et al., 2001a), 6 (Holcomb et al., 1999), or 9 (Holcomb et al., 1999) months. In the interpolated probe paradigm, APP_{695}SWE+*PS1*/M146L bigenics were impaired at 5 months (Westerman et al., 2002), but in the standard probe paradigm, they were not impaired at 5 or 15 months (Arendash et al., 2001a,b). Their performance in the visible platform subtask was equivalent to that of non-transgenic controls at 5 and 15 months (Arendash et al., 2001a). Variable results are also reported in the similar APP_{695}SWE+*PS1*/M146V bigenic, impaired during acquisition and probe testing as early as 6 months of age (Trinchese et al., 2004), but yet in separate studies not impaired during acquisition at 6 (Olcese et al., 2009) or 8 (Cracchiolo et al., 2007) months.

One obtains a clearer view concerning the impact of *APP* transgenesis by analyzing the behavior of other mutants such as APP_{751}SWE (APP23) on the behaviorally efficient

C57BL/6J background strain. Indeed, the acquisition of the hidden platform subtask was slowed down in *APP*$_{751}$SWE transgenic mutants from 16 to 24 months of age (Dumont et al., 2004; Kelly et al., 2003; Lalonde et al., 2002b; Sykova et al., 2005; Van Dijck et al., 2009), even prior to plaque deposition at 3 (Huang et al., 2006; Kelly et al., 2003; Van Dam et al., 2003) or 4 (Kawahara et al., 2012) months, probably due to high concentrations of soluble Aβ or Aβ oligomers. In contrast, the mutants behaved normally in the visible platform task (Dumont et al., 2004; Huang et al., 2006; Kelly et al., 2003; Lalonde et al., 2002b). Results are more variable during the probe trial, impairments being seen in mixed-gender 3- and 18-month-old (Kelly et al., 2003), male 3- and 6-month-old (Van Dam et al., 2003), and male 12-month-old (Van Dijck et al., 2009) mice, but not in female 16- to 24-month-old mice (Dumont et al., 2004; Lalonde et al., 2002b). In addition to sex- or age-related factors, this variability may be caused by length of training.

Deficits in hidden platform and probe tests were further documented in *APP*$_{695}$SWE/*Thy1* mutants before (6 months) or after (12, 18, and 24 months) Aβ plaque deposition (Richardson et al., 2003). The acquisition of the hidden platform task was also slowed down in *APP*$_{695}$SWEch mice without Aβ plaques, though the effect was limited to one female non-congenic line as well as the probe trial in the same non-congenic line for both sexes (Savonenko et al., 2003). All groups performed normally in the visible platform condition. In a second series of tests, results were identical with respect to the hidden platform task, female transgenics being affected (Jankowsky et al., 2005). In *APP*$_{695}$SWEch+*PS1*/A246E bigenics containing Aβ plaques, place learning, probe trial, and visuomotor performance were impaired in most instances relative to controls (Liu et al., 2002; Puoliväli et al., 2002; Wang et al., 2010; Zhang et al., 2013).

The acquisition of the hidden platform task was retarded in yet another bigenic, *APP*$_{751}$SL+*PS1*/M146L (Blanchard et al., 2009).

Moreover, deficits in place learning and the probe trial were reported in six other models containing two or more mutations and Aβ plaques, namely *APP*$_{695}$SWEch+*PS1*/ΔE9 (Jankowsky et al., 2005; Savonenko et al., 2005), *APP*$_{695}$SWE/co+*PS1*/ΔE9 (Bernardo et al., 2009; Chen et al., 2012; Dinamarca et al., 2008; Ding et al., 2008; Frye and Walf, 2008; Ferguson et al., 2013; Gimbel et al., 2010; Harrison et al., 2009a,b; Hoojimans et al., 2009; Kemppainen et al., 2012; Lalonde et al., 2005; Malm et al., 2007; Minkevicienne et al., 2008; O'Reilly and Lynch, 2012; Stover and Brown, 2012; Yoshiike et al., 2008; Zhang et al., 2012), *APP*swe+*PS1*/L166P (Lo et al., 2013), 3xTg-AD (Billings et al., 2005; Blurton-Jones et al., 2009; Caccamo et al., 2010; Clinton et al., 2007; Corona et al., 2010; Giménez-Llort et al., 2007; Gulinello et al., 2009; Halagappa et al., 2007; Martinez-Coria et al., 2010; McKee et al., 2008; Nelson et al., 2007), PLB1 (Platt et al., 2011), and 5xFAD (Hongpaisan et al., 2011; Kalinin et al., 2012; Ohno et al., 2006), though variable results are found for certain mutants at a young age. For example, *APP*$_{695}$SWE/co+*PS1*/ΔE9 mice were impaired during acquisition and the probe trial at 3 months of age, prior to Aβ plaque formation (Zhang et al., 2012), yet acquisition was normal at 7 months (Su et al., 2012).

Despite APP accumulation in retina (Dutescu et al., 2009), these mutants performed normally in the visible platform test (Bernardo et al., 2009; Gimbel et al., 2010; Harrison et al., 2009a; Lalonde et al., 2005). Five null results regarding acquisition and the probe trial have appeared in 3xTg-AD mice at an age when previous experiments detected deficits (Marlatt et al., 2013; Masciopinto et al., 2012; Peng et al., 2012; Pietropaolo et al., 2008,

2009), probably because of different genetic backgrounds used from one experiment to the next as unspecified genes interact with the functional impact of amyloid deposition.

Table 2. Hidden platform, standard probe trial, and visible platform subtasks in the Morris water maze by transgenic mice expressing human mutated or wild-type (WT) *APP* genes with or without mutated *PS1*

Name	Hidden	Probe	Visible	References
APP$_{695}$SWE (Tg2576)	↓	↓	↓	Adriani et al., 2006; Arendash et al., 2004; Bianchi et al., 2008; Fernandez-Fernandez et al., 2012; Fukumoto et al., 2010; Hongpaisan et al., 2011; Hsiao et al., 1996; King and Arendash, 2002b; Lassalle et al., 2008; Lee et al., 2006a, 2011; Li et al., 2003; Lindner et al., 2006; Massaad et al., 2009; Nishida et al., 2006; Quinn et al., 2007; Ribé et al., 2005; Ribes et al., 2008; Ricobaraza et al., 2009; Westerman et al., 2002
APP$_{695}$SWEch	↓	↓	-	Jankowsky et al., 2005; Savonenko et al., 2003, 2005
APP$_{695}$SWE/*Thy1*	↓	↓	?	Richardson et al., 2003
APP$_{751}$SWE (APP23)	↓	↓	-	Dumont et al., 2004; Huang et al., 2006; Kawahara et al., 2012; Kelly et al., 2003; Lalonde et al., 2002b; Sykova et al., 2005; Van Dam et al., 2003; Van Dijck et al., 2009
APP/IND (PDAPP)	↓	↓	↓	Brody and Holtzman, 2006; Chen et al., 2000; Daumas et al., 2008; Nilsson et al., 2004
APP$_{695}$SWE/IND (TgCRND8 or Tg19959)	↓	↓	↓	Chishti et al., 2001; Dumont et al., 2009, 2010; Durairajan et al., 2012; Glazner et al., 2010; Hanna et al., 2009; Hyde et al., 2005; Janus, 2004; Janus et al., 2000; Tampellini et al., 2010
APP/SWE/IND/ J9	↓	↓	↓	Chin et al., 2005; Dziewczapolski et al., 2009; Espana et al., 2010; Spencer et al., 2008
APP/SWE/IND/ J20	↓	↓	↓	Aboulkassim et al., 2011; Cissé et al., 2011; delpolyi et al., 2008; Escribano et al., 2010; Galvan et al., 2006, 2008; Nicolakakis et al., 2008; Palop et al., 2003; Poirier et al., 2006
APP$_{751}$SWE/IND	↓	-	?	Lee et al., 2004
APP/ArcSwe	↓	-	?	Lord et al., 2009
APP$_{695}$LD	↓	↓	-	Moechars et al., 1999; Perez-Cruz et al., 2011
APP$_{751}$LD	↓	↓	?	Zhang et al., 2006
APP/SweLd	↓	?	?	Hashimoto et al., 2010
APP/OSA	↓	↓	?	Umeda et al., 2012
APP/DU/*THY1*	normal	?	?	Gandy et al., 2010
APP$_{695}$WT/*Mt2*	↓	-	?	Yamaguchi et al., 1991
APP$_{695}$WT/*Thy1*	↓	↓	-	Moechars et al., 1999
APP$_{751}$WT	↓	↓	-	Koistinaho et al., 2001; Moran et al., 1995
APP$_{751}$WT	↓	-	?	D'Hooge et al., 1996
APP-WT/*PDGF*β	↓	↓	?	Simón et al., 2009

Name	Hidden	Probe	Visible	References
APP/C104	↓	not tested	-	Nalbantoglu et al., 1997
APP/C100-Flag	↓	↓	↓	Berger-Sweeney et al., 1999
APP/C99	↓	normal	-	Lalonde et al., 2002a
APP/C99/IND	↓	↓	?	Lee et al., 2006b
APP_{695}SWE + *PS1*/M146L (line 5.1)	↓	↓	-	Arendash et al., 2001a; Sadowski et al., 2004
APP_{695}SWE + *PS1*/M146V (line 6.2)	↓	↓	?	Trinchese et al., 2004
APP_{695}SWEch + *PS1*/A246E	↓	↓	↓	Liu et al., 2002; Puoliväli et al., 2002; Wang et al., 2010; Zhang et al., 2013
APP_{695}SWE/co + *PS1*/ΔE9	↓	↓	-	Bernardo et al., 2009; Chen et al., 2012; Dinamarca et al., 2008; Ding et al., 2008; Ferguson et al., 2013; Frye and Walf, 2008; Gimbel et al., 2010; Harrison et al., 2009a,b; Hoojimans et al., 2009; Kemppainen et al., 2012; Lalonde et al., 2005; Malm et al., 2007; Minkevicienne et al., 2008; O'Reilly and Lynch, 2012; Stover and Brown, 2012; Yoshiike et al., 2008; Zhang et al., 2012
APP_{695}SWEch + *PS1*/ΔE9	↓	↓	?	Jankowsky et al., 2005; Savonenko et al., 2005
APP_{751}SL + *PS1*/M146L	↓	?	?	Blanchard et al., 2009
*APP*swe+*PS1*/L166P	↓	↓	?	Lo et al., 2013
3x-Tg-AD	↓	↓	-	Billings et al., 2005; Blurton-Jones et al., 2009; Caccamo et al., 2010; Clinton et al., 2007; Corona et al., 2010; Giménez-Llort et al., 2007; Gulinello et al., 2009; Halagappa et al., 2007; Martinez-Coria et al., 2010; McKee et al., 2008; Nelson et al., 2007
PLB1	↓	?	-	Platt et al., 2011
5xFAD	↓	↓	?	Hongpaisan et al., 2011; Kalinin et al., 2012; Ohno et al., 2006

↓ deficit, ? untested, - no difference vs controls; WT wild-type.

2.2. APP Mice with Indiana, London, Osaka, Swedish/Indiana, Swedish/London, Swedish/Arctic, and Dutch Mutations

Like some models with Swedish mutations, place learning, probe trial, and visuomotor guidance were impaired in *APP*/IND (PDAPP) mice with the V717F Indiana mutation before and after Aβ plaque formation (Brody and Holtzman, 2006; Chen et al., 2000; Daumas et al., 2008; Nilsson et al., 2004). The acquisition of the hidden platform task was likewise slowed down in four transgenic models with Swedish/Indiana mutations: APP_{695}SWE/IND (Chishti et al., 2001; Dumont et al., 2009, 2010, 2011; Durairajan et al., 2012; Glazner et al., 2010; Hanna et al., 2009; Hyde et al., 2005; Janus, 2004; Janus et al., 2000), *APP*/SWE/IND/J9

(Dziewczapolski et al., 2009; Espana et al., 2010; Spencer et al., 2008), *APP*/SWE/IND/J20 (Aboulkassim et al., 2011; Cissé et al., 2011; delpolyi et al., 2008; Escribano et al., 2010; Galvan et al., 2006, 2008; Nicolakakis et al., 2008; Palop et al., 2003; Poirier et al., 2006), and APP_{751}SWE/IND (Lee et al., 2004). This occurred despite the absence of Aβ plaques in APP_{751}SWE/IND (Lee et al., 2004) and before their appearance in APP_{695}SWE/IND (Hyde et al., 2005). In addition, the probe trial was impaired in APP_{695}SWE/IND (Chishti et al., 2001; Dumont et al., 2009, 2010; Hyde et al., 2005; Janus, 2004; Tampellini et al., 2010), *APP*/SWE/IND/J9 (Chin et al., 2005; Dziewczapolski et al., 2009; Espana et al., 2010; Spencer et al., 2008), and *APP*/SWE/IND/J20 (Escribano et al., 2010; Galvan et al., 2006; Nicolakakis et al., 2008; Palop et al., 2003; Poirier et al., 2006). Visual platform performance was deficient in 10-month-old APP_{695}SWE/IND mice (Hyde et al., 2005) but not in younger ones (Chishti et al., 2001; Dumont et al., 2009, 2010; Hyde et al., 2005; Janus, 2004; Tampellini et al., 2010). Likewise, visual platform performance was affected in the *APP*/SWE/IND/J20 strain with high (Palop et al., 2003) but to a lesser extent in the *APP*/SWE/IND/J9 one with lower amyloid load (Chin et al., 2005; Dziewczapolski et al., 2009; Spencer et al., 2008).

The sensitivity of the hidden platform task is once again evident with other mutations carrying Aβ plaques such as *APP*/OSA mice with the E22Δ Osaka mutation (Umeda et al., 2012), two models with the V717I London mutation (APP_{695}LD and APP_{751}LD) (Moechars et al., 1999; Perez-Cruz et al., 2011; Zhang et al., 2006), *APP*/ArcSwe mice with Swedish and E693G Arctic mutations (Lord et al., 2009), and *APP*/SweLd mice with Swedish/London mutations (Hashimoto et al., 2010). Moroever, probe trial performance was impaired in APP_{695}LD (Moechars et al., 1999) and APP_{751}LD (Zhang et al., 2006) mutants, though visuomotor performance in the former was not. In contrast, *APP*/DU/*THY1* mutants characterized by cerebral amyloid angiopathy without parenchymal Aβ plaques showed normal acquisition of the submerged platform task (Gandy et al., 2010).

2.3. APP Wild-Type (WT) Mice

Despite the absence of Aβ plaques, place learning and the probe trial were impaired in APP_{695}WT/*Thy1* mice, though not visible platform performance (Moechars et al., 1999). The same pattern was discerned for all three tests in APP_{751}WT mice (Koistinaho et al., 2001; Moran et al., 1995) and in the first two in *APP*-WT/*PDGFβ* mice (Simón et al., 2009). A different APP_{751}WT model was impaired relative to controls in place learning but not in the probe trial (D'Hooge et al., 1996), as seen in APP_{695}WT/*Mt2* mice (Yamaguchi et al., 1991). Thus, mice with wild-type *APP* mutations without Aβ plaques are susceptible to similar behavioral patterns as *APP* mutants with Aβ plaques.

2.4. APP Mice with C-Terminal Fragments

A place learning deficit was reported in mice expressing the *APP*/C104 transgene with Aβ plaques (Nalbantoglu et al., 1997). In contrast, visible platform performance was intact. Place learning, as well as the probe trial, was also deficient in *APP*/C100-Flag mice without

Aβ plaques, but in these the visible platform task was impaired (Berger-Sweeney et al., 1999). Likewise, *APP*/C99/IND mice without Aβ plaques were impaired in spatial learning and the probe trial (Lee et al., 2006b). Although place learning was impaired in *APP*/C99 mice without Aβ plaques, probe trial and visuomotor guidance were not (Lalonde et al., 2002a). Thus, in every case, place learning was impaired, though results were more variable in the other two tests, indicating that C-terminal fragments cause similar damage to Aβ plaques.

2.5. APP Mice with a Null Mutation

Place learning was impaired in $APP^{\Delta/\Delta}$ mice with the exon 2 null mutation (Müller et al., 1994; Tremml et al., 1998; Wolfer et al., 1997). The $APP^{\Delta/\Delta}$ mice were also deficient in the probe trial (Müller et al., 1994; Wolfer et al., 1997), but this effect disappeared after postnatal handling, indicating the possible influence of stress-related factors (Tremml et al., 1998). The $APP^{\Delta/\Delta}$ mice performed less well than controls in initial but not later trials of the visible platform test (Müller et al., 1994). Thus, limited experimentation indicates that place learning may be affected by underexpressing as well as overexpressing *APP*.

3. Hole-Board

The Barnes circular maze consists of a brightly lit board with multiple holes where nocturnal animals are motivated to escape into the dark (chapter 2). As with the previous task, the Barnes maze requires a win-stay strategy and reference memory, since the correct response remains the same throughout training. The *APP*/IND mutants committed more errors than non-mutants during acquisition of the Barnes hole-board task, though only after plaque formation (Huitron-Resendiz et al., 2002). Five other mutants containing Aβ plaques were deficient in this task: *APP*/SWE/IND (Yamada et al., 2011), APP_{695}SWE/IND (Walker et al., 2011), *APP*/SWE/DI (Wilcock et al., 2008), APP_{695}SWE+*PS1*/M146V (Olcese et al., 2009), and APP_{695}SWE/co+*PS1*/ΔE9 (Bernardo et al., 2009; Hoojimans et al., 2009; O'Leary and Brown, 2009; Reiserer et al., 2007). In the APP_{695}SWE/co+*PS1*/ΔE9 model, no impairment occurred when an appetitive version was used which contained a landmark near the goal (Cheng et al., 2013; Pillay et al., 2008). Unlike the six mutants cited above, the acquisition of the Barnes circular maze task was normal in two other mutants with Aβ plaques: APP_{695}SWE (Arendash et al., 2004; King and Arendash, 2002b) and APP_{695}SWE+*PS1*/M146L (Arendash et al., 2001a), although in the former reversal training was impaired (Pompl et al., 1999). These two models may be less affected because of the retinal degeneration inherent on the SJL background.

4. Object Displacement

In the object displacement test, animals explore objects in a specific configuration and then at least one of them is shifted to another place, which enhances exploration of the moved object relative to stationary ones (chapter 2). *APP*$_{695}$SWE mutants have often been shown to be deficient in this task (Good et al., 2007; Hale and Good, 2005; Middei et al., 2006; Ognibene et al., 2005). The same deficit was discovered in three other mutants bearing Aβ plaques: *APP*$_{695}$SWE/co+*PS1*/ΔE9 (Frye and Walf, 2008), PLB1 (Platt et al., 2011), and 3xTg-AD (Blurton-Jones et al., 2009; Gulinello et al., 2009; Masciopinto et al., 2012).

5. Left-Right Discrimination

APP$_{695}$SWE mice were slower in acquiring a food-motivated left-right discrimination task (Deacon et al., 2008). Paradoxically, acquisition of the same task was normal in *APP*$_{695}$SWEch+*PS1*/A246E bigenics (Liu et al., 2002; Puoliväli et al., 2002). In the cross-maze version of spatial discrimination whereby left or right responses are rewarded from two starting points (chapter 2), *APP*$_{695}$SWE mice were paradoxically better than controls (Middei et al., 2004, 2006), attributable to a perseverative bias, as shown in the spontaneous alternation test (Lalonde et al., 2012). Left-right discrimination learning was also intact in *APP*$_{695}$SWEch+*PS1*/A246E bigenics exposed to a water escape paradigm (Filali and Lalonde, 2009; Filali et al., 2009), though their reversal learning was impaired (Filali and Lalonde, 2009; Filali et al., 2009), the same pattern discerned in 3xTg-AD mutants (Filali et al., 2012). Likewise, *APP*swe+*PS1*/L166P bigenics were impaired in reversing but not acquiring an adaptation of the task whereby one of 4 maze arms was baited instead of the usual two (Radde et al., 2006).

The overall results seem to indicate that spatial discrimination learning is only occasionally sensitive to *APP* transgenesis. However, reversal training, implying mental flexibility whereby correct/incorrect responses are reversed, may be more sensitive than the original left-right training.

6. Radial-Arm Maze

In the radial-arm maze task, animals are trained to find food pellets in maze arms radiating from a central platform without returning to the same one twice (chapter 2). In contrast to previous tasks, the radial-arm maze task reflects trial-dependent working memory based on a win-shift strategy.

Relative to controls, acquisition of this task was slowed down in four models carrying Aβ plaques, namely *APP*$_{695}$SWE (Hellweg et al., 2006; Scholtzova et al., 2009; Sigurdsson et al., 2004), *APP*/IND (Dodart et al., 1999), *APP*$_{695}$SWE+*PS1*/M146L (Goñi et al., 2010), and *APP*$_{751}$SWE+*PS1*/G384A (Busche et al., 2008). On the contrary, the *APP*$_{695}$SWEch model without Aβ plaques was not slowed down (Savonenko et al., 2003).

A water maze version has been designed with different starting points and a goal arm kept in the same place for the day's training but changing from one day to the next. Deficits in this version were found in several models with Aβ plaques, namely *APP*$_{695}$SWE (Karlnoski et al., 2008; Wilcock et al., 2004), *APP*$_{695}$SWE+*PS1*/M146L (Arendash et al., 2001a; Cao et al., 2009; Gordon et al., 2001), *APP*$_{695}$SWE+*PS1*/M146V (Cracchiolo et al., 2007; Kiyota et al., 2011, 2012; Olcese et al., 2009; Trinchese et al., 2004), *APP*$_{751}$SWE (Vloeberghs et al., 2006), *APP*/SWE/DI (Wilcock et al., 2008), APP_{695}SWE/IND (Lovasic et al., 2005), *APP*/SWE/IND/J20 (Du et al., 2011; Fang et al., 2012; Yao et al., 2011), APP_{695}SWEch+*PS1*/ΔE9 (Jankowsky et al., 2005; Savonenko et al., 2005), and APP_{695}SWE/co+*PS1*/ΔE9 (Park et al., 2006) but also in the APP_{695}SWEch mutant without Aβ plaques (Jankowsky et al., 2005). The overall results indicate a high degree of sensitivity of this task to *APP* transgenesis.

7. Reinforced Left-Right Alternation

T- or Y-maze left-right alternation is a simplified version of the radial-arm maze task consisting of only two choices instead of multiple ones (chapter 2). APP_{695}SWE mice were slower to acquire food-motivated alternation at 12 months of age (Chapman et al., 1999). These results were reproduced at 8 (Barnes et al., 2004), 14 (Hale and Good, 2005), and 16 (Corcoran et al., 2002) months. However, there are two null results, one at 9 months of age when Aβ plaques start to become apparent (Deacon et al., 2008) and the other well before at 2 months (Chapman et al., 1999). Moreover, no effect was found at 6 and 14 months of age in the non-matching-to-place task (Zhuo et al., 2007).

Acknowledgments

This study was supported in part by the Alzheimer's Association of America and the National Institutes of Health. We thank Karen Minter for assistance in preparing the text.

References

Aboulkassim, T; Tong, XK; Tse, YC; Wong, TP; Woo, SB; Neet, KE; Brahimi, F; Hamel, E; Saragovi, HU. Ligand-dependent TrkA activity in brain differentially affects spatial learning and long-term memory. *Mol. Pharmacol.,* 2011, 80, 498-508.

Adriani, W; Ognibene, E; Heuland, E; Ghirardi, O; Caprioli, A; Laviola, G. Motor impulsivity in APP-SWE mice: a model of Alzheimer's disease. *Behav. Pharmacol.,* 2006, 17, 525-33.

Andrä, K; Abramowski, D; Duke, M; Probst, A; Wiederhold, KH; Bürki, K; Goedert, M; Sommer, B; Staufenbiel, M. Expression of APP in transgenic mice: a comparison of neuron-specific promoters. *Neurobiol. Aging,* 1996, 17, 183-90.

Arendash, GW; King, DL; Gordon, MN; Morgan, D; Hatcher, JM; Hope, CE; Diamond, DM. Progressive, age-related behavioral impairments in transgenic mice carrying both mutant amyloid precursor protein and presenilin-1 transgenes. *Brain Res.,* 2001a, 891, 42-53.

Arendash, GW; Lewis, J; Leighty, RE; McGowan, E; Cracchiolo, JR; Hutton, M; Garcia, MF. Multi-metric behavioral comparison of APPsw and P301L models for Alzheimer's disease: linkage of poorer cognitive performance to tau pathology in forebrain. *Brain Res.,* 2004, 1012, 29-41.

Arendash, GW; Gordon, MN; Diamond, DM; Austin, LA; Hatcher, JM; DiCarlo, G; Wilcock, D; Morgan, D. Behavioral assessment of Alzheimer's transgenic mice following long-term Aβ vaccination: task specificity and correlations between Aβ and spatial memory. *DNA Cell Biol.,* 2001b, 20, 737-44.

Barnes, P; Hale, G; Good, M. Related intramaze and extramaze cue processing in adult APPSWE Tg2576 transgenic mice. *Behav. Neurosci.,* 2004, 118, 1184-95.

Berger-Sweeney, J; McPhie, DL; Arters, JA; Greenan, J; Oster-Granite, ML; Neve, RL. Impairments in learning and memory accompanied by neurodegeneration in mice transgenic for the carboxyl-terminus of the amyloid precursor protein. *Mol. Brain Res.,* 1999, 66, 150-62.

Bernardo, A; Harrison, FE; McCord, M; Zhao, J; Bruchey, A; Davies, SS; Jackson Roberts, L 2nd; Mathews, PM; Matsuoka, Y; Ariga, T; Yu, RK; Thompson, R; McDonald, MP. Elimination of GD3 synthase improves memory and reduces amyloid-ß plaque load in transgenic mice. *Neurobiol. Aging,* 2009, 30, 1777-91.

Bianchi, SL; Tran, T; Liu, C; Lin, S; Li, Y; Keller, JM; Eckenhoff, RG; Eckenhoff, M.F. Brain and behavior changes in 12-month-old Tg2576 and nontransgenic mice exposed to anesthetics. *Neurobiol. Aging,* 2008, 29, 1002-10.

Billings, LM; Oddo, S; Green, KN; McGaugh, JL; LaFerla, FM. Intraneuronal Aβ causes the onset of early Alzheimer's disease-related cognitive deficits in transgenic mice. *Neuron,* 2005, 45, 675-88.

Blanchard, V; Decorte, L; Noguès, X; Micheau, J. Characterization of cognition alteration across the course of the disease in APP751SL mice with parallel estimation of cerebral Abeta deposition. *Behav. Brain Res.,* 2009, 201, 147-57.

Blanchard, V; Moussaoui, S; Czech, C; Touchet, N; Bonici, B; Planche, M; Canton, T; Jedidi, I; Gohin, M; Wirths, O; Bayer, T.A; Langui, D; Duyckaerts, C; Tremp, G; Pradier, L. Time sequence of maturation of dystrophic neurites associated with Aβ deposits in APP/PS1 transgenic mice. *Exp. Neurol.,* 2003, 184, 247-63.

Blurton-Jones, M; Kitazawa, M; Martinez-Coria, HA; Müller, FJ; Loring, JF; Yamasaki, TR; Poon, WW; Green, KN; LaFerla, FM. Neural stem cells improve cognition via BDNF in a transgenic model of Alzheimer disease. *Proc. Natl. Acad. Sci. USA,* 2009, 106, 13594-9.

Borchelt, D.R; Ratovitski, T; van Lare, J; Lee, MK; Gonzales, V; Jenkins, NA; Copeland, NG; Price, DL; Sisodia, SS. Accelerated amyloid deposition in the brains of transgenic mice coexpressing mutant presenilin 1 and amyloid precursor proteins. *Neuron,* 1997, 19, 939-45.

Borchelt, DR; Thinakaran, G; Eckman, CB; Lee, MK; Davenport, F; Ratovitsky, T; Prada, CM; Kim, G; Seekins, S; Yager, D; Slunt, HH; Wang, R; Seeger, M; Levey, AI; Gandy, SE; Copeland, NG; Jenkins, NA; Price, DL; Younkin, SG; Sisodia, SS. Familial Alzheimer's disease-linked presenilin 1 variants elevate Aβ1-42/1-40 ratio *in vitro* and *in vivo*. *Neuron,* 1996, 17, 1005-13.

Brody, DL; Holtzman, D.M. Morris water maze search strategy analysis in PDAPP mice before and after experimental traumatic brain injury. *Exp. Neurol.,* 2006, 197, 330-40.

Bromley-Brits, K; Deng, Y; Song, W. Morris water maze test for learning and memory deficits in Alzheimer's disease model mice. *J. Vis. Exp.,* 2011, 20, 53.

Busche, MA; Eichhoff, G; Adelsberger, H; Abramowski, D; Wiederhold, KH; Haass, C; Staufenbiel, M; Konnerth, A; Garaschuk, O. Clusters of hyperactive neurons near amyloid plaques in a mouse model of Alzheimer's disease. *Science,* 2008, 321, 1686-9.

Caccamo, A; Maldonado, MA; Bokov, AF; Majumder, S; Oddo, S. CBP gene transfer increases BDNF levels and ameliorates learning and memory deficits in a mouse model of Alzheimer's disease. *Proc. Natl. Acad. Sci. USA,* 2010, 107, 22687-92.

Cao, C; Arendash, GW; Dickson, A; Mamcarz, MB; Lin, X; Ethell, DW. Aβ-specific Th2 cells provide cognitive and pathological benefits to Alzheimer's mice without infiltrating the CNS. *Neurobiol. Dis.,* 2009, 34, 63-70.

Casas, C; Sergeant, N; Itier, JM; Blanchard, V; Wirths, O; van der Kolk, N; Vingtdeux, V; van de Steeg, E; Ret, G; Canton, T; Drobecq, H; Clark, A; Bonici, B; Delacourte, A; Benavides, J; Schmitz, C; Tremp, G; Bayer, TA; Benoit, P; Pradier, L. Massive CA1/2 neuronal loss with intraneuronal and N-terminal truncated $A\beta_{42}$ accumulation in a novel Alzheimer transgenic model. *Am. J. Pathol.*, 2004, 165, 1289-300.

Chapman, PF; White, GL; Jones, MW; Cooper-Blacketer, D; Marshall, VJ; Irizarry, M; Younkin, L; Good, MA; Bliss, TVP; Hyman, BT; Younkin, SG; Hsiao, KK. Impaired synaptic plasticity and learning in aged amyloid precursor protein transgenic mice. *Nature Neurosci.,* 1999, 2, 271-76.

Chen, G; Chen, KS; Knox, J; Inglis, J; Bernard, A; Martin, SJ; Justice, A; McConlogue, L; Games, D; Freedman, SB; Morris, RGM. A learning deficit related to age and β-amyloid plaques in a mouse model of Alzheimer's disease. *Nature,* 2000, 408, 975-9.

Chen, SQ; Cai, Q; Shen, YY; Wang, PJ; Teng, GJ; Zhang, W; Zang, FC. Age-related changes in brain metabolites and cognitive function in APP/PS1 transgenic mice. *Behav. Brain Res,* 2012, 235, 1-6.

Cheng, D; Logge W; Low, JK; Garner, B; Karl, T. Novel behavioural characteristics of the APP(Swe)/PS1ΔE9 transgenic mouse model of Alzheimer's disease. *Behav. Brain Res.,* 2013, 245, 120-7.

Chin, J; Palop, JJ; Puolivali, J; Massaro, C; Bien-Ly, N; Gerstein, H; Scearce-Levie, K; Masliah, E; Mucke, L. Fyn kinase induces synaptic and cognitive impairments in a transgenic mouse model of Alzheimer's disease. *J. Neurosci.,* 2005, 25, 9694-703.

Chishti, MA; Yang, D-S; Janus, C; Phinney, AL; Horne, P; Pearson, J; Strome, R; Zuker, N; Loukides, J; French, J; Turner, S; Lozza, G; Grilli, M; Kunicki, S; Morissette, C; Paquette, J; Gervais, F; Bergeron, C; Fraser, PE; Carlson, GA; St. George-Hyslop, P; Westaway, D. Early-onset amyloid deposition and cognitive deficits in transgenic mice expressing a double mutant form of amyloid protein 695. *J. Biol. Chem.,* 2001, 276, 21562-70.

Choi, DY; Lee, JW; Peng, J; Lee, YJ; Han, JY; Lee, YH; Choi, IS; Han, SB; Jung, JK; Lee, WS; Lee, SH; Kwon, BM; Oh, KW; Hong, JT. Obovatol improves cognitive functions in animal models for Alzheimer's disease. *J. Neurochem.,* 2012, 120, 1048-59.

Cissé, M; Halabisky, B; Harris, J; Devidze, N; Dubal, DB; Sun, B; Orr, A; Lotz, G; Kim, DH; Hamto, P; Ho, K; Yu, GQ; Mucke, L. Reversing EphB2 depletion rescues cognitive functions in Alzheimer model. *Nature,* 2011, 469, 47-52.

Clinton, LK; Billings, LM; Green, KN; Caccamo, A; Ngo, J; Oddo, S., McGaugh, JL; LaFerla, FM. Age-dependent sexual dimorphism in cognition and stress response in the 3xTg-AD mice. *Neurobiol. Dis.,* 2007, 28, 76-82.

Cole, GM; Frautschy, SA. Animal models for Alzheimer's disease. *Alz. Dis. Rev.,* 1997, 2, 33-41.

Corcoran, KA; Lu, Y; Turner, RS; Maren, S. Overexpression of *hAPPswe* impairs rewarded alternation and contextual fear conditioning in a transgenic mouse model of Alzheimer's disease. *Learn. Mem.,* 2002, 9, 243-52.

Corona, C; Masciopinto, F; Silvestri, E; Viscovo, AD; Lattanzio, R; Sorda, RL; Ciavardelli, D; Goglia, F; Piantelli, M; Canzoniero, LM; Sensi, SL. Dietary zinc supplementation of 3xTg-AD mice increases BDNF levels and prevents cognitive deficits as well as mitochondrial dysfunction. *Cell. Death Dis.,* 2010, 1, e91.

Cracchiolo, JR; Mori, T; Nazian, SJ; Tan, J; Potter, H; Arendash, GW. Enhanced cognitive activity- over and above social or physical activity- is required to protect Alzheimer's mice against cognitive impairment, reduce Aβ deposition, and increase synaptic immunoreactivity. *Neurobiol. Learn. Mem.,* 2007, 88, 277-94.

Daumas, S; Sandin, J; Chen, K.S; Kobayashi, D; Tulloch, J; Martin, SJ; Games, D; Morris, RG. Faster forgetting contributes to impaired spatial memory in the PDAPP mouse: deficit in memory retrieval associated with increased sensitivity to interference? *Learn. Mem.,* 2008, 15, 625-32.

Deacon, RMJ; Cholerton, LL; Talbot, K; Nair-Roberts, RG; Sanderson, DJ; Romberg, C; Koros, E; Bornemann, KD; Rawlins, JNP. Age-dependent and -independent behavioral deficits in Tg2576 mice. *Behav. Brain Res.,* 2008, 189, 126-38.

delpolyi, AR; Fang, S; Palop, JJ; Yu, GQ; Wang, X; Mucke, L. Altered navigational strategy use and visuospatial deficits in hAPP transgenic mice. *Neurobiol. Aging,* 2008, 29, 253-66.

D'Hooge, R; Nagels, G; Mucke, L; De Deyn, PP. Spatial learning deficit in mice expressing human 751-amino acid β-amyloid precursor protein. *NeuroReport,* 1996, 7, 2807-11.

Dinamarca, MC; Arrázola, M; Toledo, E; Cerpa, WF; Hancke, J; Inestrosa, NC. Release of acetylcholinesterase (AChE) from beta-amyloid plaques assemblies improves the spatial memory impairments in APP-transgenic mice. *Chem. Biol. Interact.,* 2008, 175, 142-9.

Dineley, KT; Xia, X; Bui, D; Sweatt, JD; Zheng, H. Accelerated plaque accumulation, associative learning deficits, and up-regulation of α7 nicotinic receptor protein in transgenic mice co-expressing mutant human presenilin 1 and amyloid precursor proteins. *J. Biol. Chem.,* 2002, 277, 22768-80.

Ding, Y; Qiao, A; Wang, Z; Goodwin, JS; Lee, ES; Block, ML; Allsbrook, M; McDonald, MP; Fan, GH. Retinoic acid attenuates β-amyloid deposition and rescues memory deficits in an Alzheimer's disease transgenic mouse model. *J. Neurosci.,* 2008, 28, 11622-34.

Dodart, J-C; Mathis, C; Bales, KR; Paul, SM. Does my mouse have Alzheimer's disease? *Genes Brain Behav.,* 2002, 1, 142-55.

Dodart, J-C; Meziane, H; Mathis, C; Bales, KR; Paul, SM; Ungerer, A. Behavioral disturbances in transgenic mice overexpressing the V717F β-amyloid precursor protein. *Behav. Neurosci.,* 1999, 113, 982-90.

Du, H; Guo, L; Zhang, W; Rydzewska, M; Yan, S. Cyclophilin D deficiency improves mitochondrial function and learning/memory in aging Alzheimer disease mouse model *Neurobiol. Aging,* 2011, 32, 398-406.

Duff, K. Transgenic mouse models of Alzheimer's disease. In: W Crusio, R Gerlai (eds) *Handbook of molecular-genetic techniques for brain and behavior research* (Techniques in the Behavioral and Neural Sciences, vol 13). Amsterdam: Elsevier, pp 888-94, 1999.

Dumont, M; Strazielle, C; Staufenbiel, M; Lalonde, R. Spatial learning and exploration of environmental stimuli in 24-month-old female APP23 transgenic mice with the Swedish mutation. *Brain Res.,* 2004, 1024, 113-21.

Dumont, M; Stack, C; Elipenhali, C; Calingasan, NY; Wille, E; Beal, MF. Apocynin administration does not improve behavioral and neuropathological deficits in a transgenic mouse model of Alzheimer's disease. *Neurosci. Lett.,* 2011, 492, 150-4.

Dumont, M; Wille, E; Calingasan, NY; Nathan, C; Beal, MF; Lin, MT. N-iminoethyl-L-lysine improves memory and reduces amyloid pathology in a transgenic mouse model of amyloid deposiiton. *Neurochem. Int.,* 2010, 56, 345-51.

Dumont, M; Wille, E; Calingasan, NY; Tampellini, D; Williams, C; Gouras, GK; Liby, K; Sporn, M; Nathan, C; Flint Beal, M; Lin, MT. Triterpenoid CDDO-methylamide improves memory and decreases amyloid plaques in a transgenic mouse model of Alzheimer's disease. *J. Neurochem.,* 2009, 109, 502-12.

Durairajan, SS; Liu, LF; Lu, JH; Chen, LL; Yuan, Q; Chung, SK; Huang, L; Li, XS; Huang, JD; Li, M. Berberine ameliorates β-amyloid pathology, gliosis, and cognitive impairment in an Alzheimer's disease transgenic mouse model. *Neurobiol. Aging*, 2012, 33, 2903-19.

Dutescu, RM; Li, QX; Crowston, J; Masters, CL; Baird, PN; Culvenor, JG. Amyloid precursor protein processing and retinal pathology in mouse models of Alzheimer's disease. *Graefes. Arch. Clin. Exp. Ophthalmol.*, 2009, 247, 1213-21.

Dziewczapolski, G; Glogowski, CM; Masliah, E; Heinemann, SF. Deletion of the α7 nicotinic acetylcholine receptor gene improves cognitive deficits and synaptic pathology in a mouse model of Alzheimer's disease. *J. Neurosci.,* 2009, 29, 8805-15.

Escribano, L; Simon, AM; Gimeno, E; Cuadrado-Tejedor, M; Lopez de Maturana, R; Garcia-Osta, A; Ricobaraza, A; Pérez-Mediavilla, A; Del Rio, J; Frechilla, D. Rosiglitazone rescues memory impairment in Alzheimer's transgenic mice: mechanisms involving a reduced amyloid and tau pathology. *Neuropsychopharmacology,* 2010, 35, 1593-604.

Espana, J; Valero, J; Minano-Molina, AJ; Masgrau, R; Martin, E; Guardia-Laguarta, C; Lleo, A; Giménez-Llort, L; Rodriguez-Alvarez, J; Saura, CA. β-Amyloid disrupts activity-dependent gene transcription required for memory through the CREB coactivator CRTC1. *J. Neurosci.,* 2010, 30, 9402-10.

Fang, F; Chen, X; Huang, T; Lue, LF; Luddy, JS; Yan, SS. Multi-faced neuroprotective effects of Ginsenoside Rg1 in an Alzheimer mouse model. *Biochim. Biophys. Acta*, 2012, 1822, 286-92.

Ferguson, SA; Sarkar, S; Schmued, LC. Longitudinal behavioral changes in the APP/PS1 transgenic Alzheimer's disease model. *Behav. Brain Res.,* 2013, 242, 125-34.

Fernandez-Fernandez, L; Comes, G; Bolea, I; Valente, T; Ruiz, J; Murtra, P; Ramirez, B; Anglés, N; Reguant, J; Morelló, JR; Boada, M; Hidalgo, J; Escorihuela, RM; Unzeta, M. LMN diet, rich in polyphenols and polyunsaturated fatty acids, improves mouse cognitive decline associated with aging and Alzheimer's disease. *Behav. Brain Res.*, 2012, 228, 261-71.

Filali, M; Lalonde, R. Age-related cognitive decline and nesting behavior in an APPswe/PS1 bigenic model of Alzheimer's disease. *Brain Res.,* 2009, 1292, 93-9.

Filali, M; Lalonde, R; Rivest, S. Cognitive and non-cognitive behaviors in an APPswe/PS1 bigenic model of Alzheimer's disease. *Genes Brain Behav.*, 2009, 8, 143-8.

Filali, M; Lalonde, R; Thériault, P; Julien, C; Calon, F; Planel, E. Cognitive and non-cognitive behaviors in the triple transgenic mouse model of Alzheimer's disease expressing mutated APP, PS1, and Mapt (3xTg-AD). *Behav. Brain Res.,* 2012, 234, 334-42.

Frye, CA; Walf, AA. Effects of progesterone administration and APPswe+PSEN1ΔE9 mutation for cognitive performance of mid-aged mice. *Neurobiol. Learn. Mem.*, 2008, 89, 17-26.

Fukuchi, K; Ho, L; Younkin, SG; Kunkel, DD; Ogburn, CE; LeBoeuf, RC; Furlong, CE; Deeb, SS; Nochlin, D; Wegiel, J; Wisniewski, H.M; Martin, GM. High levels of circulating β-amyloid peptide do not cause cerebral β-amyloidosis in transgenic mice. *Am. J. Pathol.,* 1996, 149, 219-27.

Fukumoto, H; Takahashi, H; Tarui, N; Matsui, J; Tomita, T; Hirode, M; Sagayama, M; Maeda, R; Kawamoto, M; Hirai, K; Terauchi, J; Sakura, Y; Kakihana, M; Kato, K; Iwatsubo, T; Miyamoto, M. A noncompetitive BACE1 inhibitor TAK-070 ameliorates Abeta pathology and behavioral deficits in a mouse model of Alzheimer's disease. *J. Neurosci.,* 2010, 30, 11157-66.

Galvan, V; Zhang, J; Gorostiza, OF; Banwait, S; Huang, W; Ataie, M; Tang, H; Bredesen, DE. Long-term prevention of Alzheimer's disease-like behavioral deficits in PDAPP mice carrying a mutation in Asp664. *Behav. Brain Res.*, 2008, 191, 246-55.

Galvan, V; Gorostiza, OF; Banwait, S; Ataie, M; Logvinova, AV; Sitaraman, S; Carlson, E; Sagi, SA; Chevallier, N; Jin, K; Greenberg, DA; Bredesen, DE. Reversal of Alzheimer's-like pathology and behavior in human APP transgenic mice by mutation of Asp664. *Proc. Natl. Acad. Sci. USA,* 2006, 103, 7130-5.

Games, D; Adams, D; Alessandri, R; Barbour, R; Berthelette, P; Blackwell, C; Carr, T; Clemens, J; Donaldson, T; Gillespie, F; Guido, T; Hagopian, S; Johnson-Wood, K; Khan, K; Lee, M; Leibowitz, P; Lieberburg, I; Little, S; Masliah, E; McConlogue, L; Montoya-Zavala, M; Mucke, L; Paganini, L; Penniman, E; Power, M; Schenk, D; Seubert, P; Snyder, B; Soriano, F; Tan, H; Vitale, J; Wadworth, S; Wolozin, B; Zhao, J. Alzheimer-type neuropathology in transgenic mice overexpressing V717F β-amyloid precursor protein. *Nature,* 1995, 373, 523-7.

Gandy, S; Simon, AJ; Steele, JW; Lublin, AL; Lah, JJ; Walker, LC; Levey, AI; Krafft, GA; Levy, E; Checler, F; Glabe, C; Bilker, WB; Abel, T; Schmeidler, J; Ehrlich, ME. Days to criterion as an indicator of toxicity associated with human Alzheimer amyloid-β oligomers. *Ann. Neurol.,* 2010, 68, 220-30.

German, DC; Eisch AJ. Mouse models of Alzheimer's disease: insight into treatment. *Rev. Neurosci.,* 2004, 15, 353-69.

Gimbel, DA; Nygaard, HB; Coffey, EE; Gunther, EC; Lauren, J; Gimbel, ZA; Strittmatter, SM. Memory impairment in transgenic Alzheimer mice requires cellular prion protein. *J. Neurosci.,* 2010, 30, 6367-74.

Giménez-Llort, L; Blasquez, G; Canete, T; Johansson, B; Oddo, S; Tobena, A; LaFerla, FM; Fernandez-Teruel, A. Modeling behavioral and neuronal symptoms of Alzheimer's disease in mice: a role for intraneuronal amyloid. *Neurosci. Biobehav. Rev.*, 2007, 31, 125-47.

Glazner, KA; Odero, GL; Anema, E; Motnenko, A; Schapansky, J; Grossman, D; Oliver, DR; Glazner, GW; Albensi, BC. Strain specific differences in memory and neuropathology in a mouse model of Alzheimer's disease. *Life Sci.,* 2010, 86, 942-50.

Goñi, F; Prelli, F; Ji Y; Scholtzova, H; Yang, J; Sun, Y; Liang, FX; Kascsak, R; Kascsak, R; Mehta, P; Wisniewski, T. Immunomodulation targeting abnormal protein conformation reduces pathology in a mouse model of Alzheimer's disease. *PLoS One,* 2010, 5(10):e13391.

Good, MA; Hale, G; Staal, V. Impaired "episodic-like" object memory in adult APPswe transgenic mice. *Behav. Neurosci.,* 2007, 121, 443-8.

Gordon, MN; King, DL; Diamond, DM; Jantzen, PT; Boyett, KV; Hope, CE; Hatcher, JM; DiCarlo, G; Gottscall, WPE; Morgan, D; Arendash, GW. Correlation between cognitive deficits and Aβ deposits in transgenic APP+PS1 mice. *Neurobiol. Aging,* 2001, 22, 377-86.

Gulinello, M; Gertner, M; Mendoza, G; Schoenfeld, BP; Oddo, S; LaFerla, F; Choi, CH; McBride, SM; Faber, DS. Validation of a 2-day water maze protocol in mice. *Behav. Brain Res.,* 2009, 196, 220-7.

Halagappa, VK; Guo, Z; Pearson, M; Matsuoka, Y; Cutler, RG; Laferla, FM; Mattson, MP. Intermittent fasting and caloric restriction ameliorate age-related behavioral deficits in the triple-transgenic mouse model of Alzheimer's disease. *Neurobiol. Dis.,* 2007, 26, 212-20.

Hale, G; Good, M. Impaired visuospatial recognition memory but normal object novelty detection and relative familiarity judgments in adult mice expressing the APPswe Alzheimer's disease mutation. *Behav. Neurosci.,* 2005, 119, 884-91.

Hanna, A; Horne, P; Yager, D; Eckman, C; Eckman, E; Janus, C. Amyloid beta and impairment in multiple memory systems in older transgenic APP TgCRND8 mice. *Genes Brain Behav.,* 2009, 8, 676-84.

Harrison, FE; Hosseini, AH; McDonald, MP; May, JM. Vitamin C reduces spatial learning deficits in middle-aged and very old APP/PSEN1 transgenic and wild-type mice. *Pharmacol. Biochem. Behav.,* 2009a, 93, 443-50.

Harrison, FE; Allard, J; Bixler, R; Usoh, C; Li, L; May, JM; McDonald, MP. Antioxidants and cognitive training interact to affect oxidative stress and memory in APP/PSEN1 mice. *Nutr. Neurosci.,* 2009b, 12, 203-18.

Hashimoto, T; Ogino, K; Shin, RW; Kitamoto, T; Kikuchi, T; Shimizu, N. Age-dependent increase in lysosome-associated membrane protein 1 and early-onset behavioral deficits in APPSL transgenic mouse model of Alzheimer's disease. *Neurosci. Lett.,* 2010, 469, 273-7.

Hellweg, R; Lohmann, P; Huber, R; Kühl, A; Riepe, MW. Spatial navigation in complex and radial mazes in APP23 animals and neurotrophin signaling as a biological marker of early impairment. *Learn. Mem.,* 2006, 13, 63-71.

Hock, C; Lamb, BT. Transgenic mouse models of Alzheimer's disease. *Trends Genet.,* 2001, 17, S7-S12.

Holcomb, LA; Gordon, MN; Jantzen, P; Hsiao, K; Duff, K; Morgan, D. Behavioral changes in transgenic mice expressing both amyloid precursor protein and presenilin-1 mutations: lack of association with amyloid deposits. *Behav. Genet.,* 1999, 29, 177-85.

Holcomb, L; Gordon, MN; McGowan, E; Yu, X; Benkowic, S; Jantzen, P; Wright, K; Saad, I; Mueller, R; Morgan, D; Sanders, S; Zehr, C; O'Campo, K; Hardy, J; Prada, C-M; Eckman, C; Younkin, S; Hsiao, K; Duff, K. Accelerated Alzheimer-type phenotype in transgenic mice carrying both mutant amyloid precursor protein and presenilin transgenes. *Nature Med.,* 1998, 4, 97-100.

Hongpaisan, J; Sun, M-K; Alkon, DL. PKCε activation prevents synaptic loss, Aβ elevation, and cognitive deficits in Alzheimer's disease transgenic mice. *J. Neurosci.,* 2011, 31, 630-43.

Hooijmans, CR; Van der Zee, CE; Dederen, PJ; Brouwer, KM; Reijmer, YD; van Groen, T; Broersen, LM; Lütjohann, D; Heerschap, A; Kiliaan, AJ. DHA and cholesterol containing diets influence Alzheimer-like pathology, cognition and cerebral vasculature in APPswe/PS1dE9 mice. *Neurobiol. Dis.,* 2009, 33, 482-98.

Howlett, DR; Richardson, JC; Austin, A; Parsons, AA; Bate, ST; Davies, DC; Gonzalez, MI. Cognitive correlates of Aβ deposition in male and female mice bearing amyloid precursor protein and presenilin-1 mutant transgenes. *Brain Res.,* 2004, 1017, 130-6.

Hsiao, K. Transgenic mice expressing Alzheimer amyloid precursor proteins. *Exp. Gerontol.,* 1998a, 33, 883-9.

Hsiao, K. Strain dependent and invariant features of transgenic mice expressing Alzheimer amyloid precursor proteins. *Prog. Brain Res.,* 1998b, 117, 335-41.

Hsiao, K; Chapman, P; Nilsen, S; Eckman, C; Harigaya, Y; Younkin, S; Yang, F; Cole, G. Correlative memory deficits, Aβ elevation, and amyloid plaques in transgenic mice. *Science*, 1996, 274, 99-102.

Huang, SM; Mouri, A; Kokubo, H; Nakajima, R; Suemoto, T; Higuchi, M; Staufenbiel, M; Noda, Y; Yamaguchi, H; Nabeshima, T; Saido, TC; Iwata, N. Neprilysin-sensitive synapse-associated amyloid-beta peptide oligomers impair neuronal plasticity and cognitive function. *J. Biol. Chem.,* 2006, 281, 17941-51.

Huitron-Resendiz, S; Sanchez-Alavez, M; Gallegos, R; Berg, G; Crawford, E; Giacchino, JL; Games, D; Henriksen, SJ; Criado, JR. Age-independent and age-related deficits in visuospatial learning, sleep-wake states, thermoregulation and motor activity in PDAPP mice. *Brain Res*., 2002, 928, 126-37.

Hyde, LA; Kazdoba, TM; Grilli, M; Lozza, G; Brussa, R; Zhang, Q; Wong, GT; McCool, MF; Zhang, L; Parker, EM; Higgins, GA. Age-progressing cognitive impairments and neuropathology in transgenic CRND8 mice. *Behav. Brain Res.,* 2005, 160, 344-55.

Jankowsky, JL; Melnikova, T; Fadale, DJ; Xu, GM; Slunt, HH; Gonzales, V; Younkin, LH; Younkin, SG; Borchelt, DR; Savonenko, AV. Environmental enrichment mitigates cognitive deficits in a mouse model of Alzheimer's disease. *J. Neurosci.,* 2005, 25, 5217-24.

Jankowsky, JL; Fadale, DJ; Anderson, J; Xu, GM; Gonzales, V; Jenkins, NA; Copeland, NG; Lee, MK; Younkin, LH; Wagner SL; Younkin, SG; Borchelt, DR. Mutant presenilins specifically elevate the levels of the 42 residue β-amyloid peptide *in vivo*: evidence for augmentation of a 42-specific γ secretase. *Hum. Mol. Genet.*, 2004, 13, 159-70.

Janus, C. Search strategies used by *APP* transgenic mice during navigation in the Morris water maze. *Learn. Mem.,* 2004, 11, 337-46.

Janus, C; Westaway, D. Transgenic mouse models of Alzheimer's disease. *Physiol. Behav.,* 2001, 73, 873-86.

Janus, C; Pearson, J; McLaurin, J; Matthews, PM; Jiang, Y; Schmidt, SD; Chishti, MA; Horne, P; Heslin, D; French, J; Mount, HTJ; Nixon, RA; Mercken, M; Bergeron, C; Fraser, PE; St George-Hyslop, P; Westaway, D. Aβ peptide immunization reduces behavioural impairment and plaques in a model of Alzheimer's disease. *Nature,* 2000, 408, 979-82.

Kalinin, S; Polak, PE; Lin, SX; Sakharkar, AJ; Pandey, SC; Feinstein, DL. The noradrenaline precursor L-DOPS reduces pathology in a mouse model of Alzheimer's disease. *Neurobiol. Aging,* 2012, 33, 1651-63.

Karl, T; Bhatia, S; Cheng, D; Kim, WS; Garner, B. Cognitive phenotyping of amyloid precursor protein transgenic J20 mice. *Behav. Brain Res.,* 2012, 228, 392-7.

Karlnoski, RA; Rosenthal, A; Alamed, J; Ronan, V; Gordon, MN; Gottschall, PE; Grimm, J; Pons, J; Morgan, D. Deglycosylated anti-Aβ antibody dose-response effects on pathology and memory in APP transgenic mice. *J. Neuroimmune Pharmacol.,* 2008, 3, 187-97.

Kawahara, K; Suenobu, M; Yoshida, A; Koga, K; Hyodo, A; Ohtsuka, H; Kuniyasu, A; Tamamaki, N; Sugimoto, Y; Nakayama, H. Intracerebral microinjection of interleukin-4/interleukin-13 reduces β-amyloid accumulation in the ipsilateral side and improves cognitive deficits in young amyloid precursor protein 23 mice. *Neuroscience,* 2012, 207, 243-60.

Kelly, PH; Bondolfi, L; Hunziker, D; Schlecht, H-P; Carver, K; Maguire, E; Abramowski, D; Wiederhold, K-H; Sturchler-Pierrat, C; Jucker, M; Bergmann, R; Staufenbiel, M; Sommer, B. Progressive age-related impairment of cognitive behavior in APP23 transgenic mice. *Neurobiol. Aging,* 2003, 24, 365-8.

Kemppainen, S; Rantamäki, T; Jerónimo-Santos, A; Lavasseur, G; Autio, H; Karpova, N; Kärkkäinen, E; Stavén, S; Miranda, HV; Outeiro, TF; Diógenes, MJ; Laroche, S; Davis, S; Sebastião, AM; Castrén, E; Tanila, H. Impaired TrkB receptor signaling contributes to memory impairment in APP/PS1 mice. *Neurobiol. Aging,* 2012, 33, 1122.e23-39.

King, DL; Arendash, GW. Maintained synaptophysin immunoreactivity in Tg2576 transgenic mice during aging: correlations with cognitive impairment. *Brain Res.,* 2002a, 926, 58-68.

King, DL; Arendash, GW. Behavioral characterization of the Tg2576 transgenic model of Alzheimer's disease through 19 months. *Physiol. Behav.,* 2002b, 75, 627-42.

Kitamoto, T; Mohri, S; Ironside, JW; Miyoshi, I; Tanaka, T; Kitamoto, N; Itohara, S; Kasai, N; Katsuki, M; Higuchi, J; Muramoto, T; Shin, RW. Follicular dendritic cell of the knock-in mouse provides a new bioassay for human prions. *Biochem. Biophys. Res. Comm.,* 2002, 294, 280-6.

Kiyota, T; Ingraham, KL; Jacobsen, MT; Xiong, H; Ikezu, T. *FGF2* gene transfer restores hippocampal functions in mouse models of Alzheimer's disease and has therapeutic implications for neurocognitive disorders. *Proc. Natl. Acad. Sci. USA,* 2011, 108, E1339-48.

Kiyota, T; Ingraham, KL; Swan, RJ; Jacobsen, MT; Andrews, SJ; Ikezu, T. AAV serotype 2/1-mediated gene delivery of anti-inflammatory interleukin-10 enhances neurogenesis and cognitive function in APP+PS1 mice. *Gene Ther.,* 2012, 19, 724-33.

Koistinaho, M; Ort, M; Cimadevilla, JM; Vondrous, R; Cordell, B; Koistihano, J; Bures, J; Higgins, LS. Specific spatial learning deficits become severe with age in β-amyloid precursor protein transgenic mice that harbor diffuse β-amyloid deposits but do not form plaques. *Proc. Natl. Acad. Sci. USA,* 2001, 98, 14675-80.

Kumar-Singh, S; Dewachter, I; Moechars, D; Lübke, U; De Jonghe, C; Ceuterick, C; Checler, F; Naidu, A; Cordell, B; Cras, P., Van Broeckhoven, C; Van Leuven, F. Behavioral disturbances without amyloid deposits in mice overexpressing human amyloid precursor protein with Flemish (A692G) or Dutch (E693Q) mutation. *Neurobiol. Dis.,* 2000, 7, 9-22.

Lalonde, R; Fukuchi, K; Strazielle, C. *APP* transgenic mice for modelling behavioural and psychological symptoms of dementia (BPSD). *Neurosci. Biobehav. Rev.,* 2012, 36, 1357-75.

Lalonde, R; Kim, H-D; Maxwell, JA; Fukuchi, K. Exploratory activity and spatial learning in 12-month-old APP_{695}SWE/co+*PS1*/ΔE9 mice with amyloid plaques. *Neurosci. Lett.,* 2005, 390, 87-92.

Lalonde, R; Dumont, M; Fukuchi, K; Strazielle, C. Transgenic mice expressing the human C99 terminal fragment of *βAPP*: effects on spatial learning, exploration, anxiety, and motor coordination. *Exp. Gerontol.,* 2002a, 37, 1399-410.

Lalonde, R; Dumont, M; Staufenbiel, M; Sturchler-Pierrat, C; Strazielle, C. Spatial learning, exploration, anxiety, and motor coordination in female APP23 transgenic mice with the Swedish mutation. *Brain Res.*, 2002b, 956, 36-44.

Lassalle, JM; Halley, H; Daumas, S; Verret, L; Francés, B. Effects of the genetic background on cognitive performances of TG2576 mice. *Behav. Brain Res.,* 2008, 191, 104-10.

Lee, EB; Leng, LZ; Zhang, B; Kwong, L; Trojanowski, JQ; Abel, T; Lee, VM. Targeting amyloid-β peptide (Aβ) oligomers by passive immunization with a conformation-selective monoclonal antibody improves learning and memory in Aβ Precursor Protein (APP) transgenic mice. *J. Biol. Chem.,* 2006a, 281, 4292-99.

Lee, JS; Im, DS; An, YS; Hong, JM; Gwag, BJ; Joo, IS. Chronic cerebral hypoperfusion in a mouse model of Alzheimer's disease: an additional contributing factor of cognitive impairment. *Neurosci. Lett.,* 2011, 489, 84-8.

Lee, K-W; Lee, SH; Kim, H; Song, JS; Yang, SD; Paik, SG; Han, PL. Progressive cognitive impairment and anxiety induction in the absence of plaque deposition in C57BL/6 inbred mice expressing transgenic amyloid precursor protein. *J. Neurosci. Res.,* 2004, 76, 572-80.

Lee, K-W; Im, JY; Song, JS; Lee, SH; Lee, HJ; Ha, HY; Koh, JY; Gwag, BJ; Yang, SD; Paik, SG; Han, PL. Progressive neuronal loss and behavioral impairments of transgenic C57BL/6 inbred mice expressing the carboxy terminus of amyloid precursor protein. *Neurobiol. Dis.,* 2006b, 22, 10-24.

Li, L; Cao, D; Garber, D.W; Kim, H; Fukuchi, K. Association of aortic atherosclerosis with cerebral beta-amyloidosis and learning deficits in a mouse model of Alzheimer's disease. *Am. J. Pathol.,* 2003, 163, 2155-64.

Lindner, MD; Hogan, JB; Krause, RG; Machet, F; Bourin, C; Hodges, DB. Jr; Corsa, JA; Barten, DM; Toyn, JH; Stock, DA; Rose, GM; Gribkoff, VK. Soluble Aβ and cognitive function in aged F-344 rats and Tg2576 mice. *Behav. Brain Res.,* 2006, 173, 62-75.

Liu, L; Ikonen, S; Heikkinen, T; Tapiola, T; van Groen, T; Tanila, H. The effects of long-term treatment with metrifonate, a cholinesterase inhibitor, on cholinergic activity, amyloid pathology, and cognitive function in APP and PS1 doubly transgenic mice. *Exp. Neurol.,* 2002 173, 196-204.

Lo, AC; Iscru, E; Blum, D; Tesseur, I; Callaerts-Vegh, Z; Buée, L; De Strooper, B; Balschun, D; D'Hooge, R. Amyloid and tau neuropathology differentially affect prefrontal synaptic plasticity and cognitive performance in mouse models of Alzheimer's disease. *J. Alzheimers Dis.,* 2013, 37, 109-25.

Lord, A; Kalimo, H; Eckman, C; Zhang, X-Q; Lannfelt, L; Nilsson, LNG. The Arctic Alzheimer mutation facilitates early intraneuronal Aβ aggregation and senile plaque formation in transgenic mice. *Neurobiol. Aging,* 2006, 27, 67-77.

Lord, A; Englund, H; Söderberg, L; Tucker, S; Clausen, F; Hillered, L; Gordon, M; Morgan, D; Lannfelt, L; Pettersson, FE; Nilsson, LN. Amyloid-β protofibril levels correlate with spatial learning in Arctic Alzheimer's disease transgenic mice. *FEBS J.,* 2009, 276, 995-1006.

Lovasic, L; Bauschke, H; Janus, C. Working memory impairment in a transgenic amyloid precursor protein TgCRND8 mouse model of Alzheimer's disease. *Genes Brain Behav.,* 2005, 4, 197-208.

Malm, TM; Iivonen, H; Goldsteins, G; Keksa-Goldsteine, V; Ahtoniemi, T; Kanninen, K; Salminen, A; Auriola, S; Van Groen, T; Tanila, H; Koistinaho, J. Pyrrolidine dithiocarbamate activates Akt and improves spatial learning in APP/PS1 mice without affecting beta-amyloid burden. *J. Neurosci.,* 2007, 27, 3712-21.

Marlatt, MW; Potter, MC; Bayer, TA; van Praag, H; Lucassen, PJ. Prolonged running, not fluoxetine treatment, increases neurogenesis, but does not alter neuropathology, in the 3xTg mouse model of Alzheimer's disease. *Curr. Topics Behav. Neurosci.,* 2013, 15, 313-40.

Martinez-Coria, H; Green, KN; Billings, LM; Kitazawa, M; Albrecht, M; Rammes, G; Parsons, CG; Gupta, S; Banerjee, P; LaFerla, FM. Memantine improves cognition and reduces Alzheimer's-like neuropathology in transgenic mice. *Am. J. Pathol.,* 2010, 176, 870-80.

Masciopinto, F; Di Pietro, N; Corona, C; Bomba, M; Pipino, C; Curcio, M; Di Castelnuovo, A; Ciavardelli, D; Silvestri, E; Canzoniero, LM; Sekler, I; Pandolfi, A; Sensi, SL. Effects of long-term treatment with pioglitazone on cognition and glucose metabolism of PS1-KI, 3xTg-AD, and wild-type mice. *Cell Death Dis.,* 2012, 3, e448.

Massaad, CA; Washington, TM; Pautler, RG; Klann, E. Overexpression of SOD-2 reduces hippocampal superoxide and prevents memory deficits in a mouse model of Alzheimer's disease. *Proc. Natl. Acad. Sci. USA,* 2009, 106, 13576-81.

McDonald, MP; Overmier, JB. Present imperfect: a critical review of animals models of the mnemonic impairments in Alzheimer's disease. *Neurosci. Biobehav. Rev.,* 1998, 22, 99-120.

McKee, AC; Carreras, I; Hossain, L; Ryu, H; Klein, WL; Oddo, S; LaFerla, FM; Jenkins, BG; Kowall, NW; Dedeoglu, A. Ibuprofen reduces Aβ, hyperphosphorylated tau and memory deficits in Alzheimer mice. *Brain Res.,* 2008, 1207, 225-36.

Melov, S; Wolf, N; Strozyk, D; Doctrow, SR; Bush, AI. Mice transgenic for Alzheimer disease β-amyloid develop lens cataracts that are rescued by antioxidant treatment. *Free Radic. Biol. Med.,* 2005, 38, 258-61.

Middei, S; Daniele, S; Caprioli, A; Ghirardi, O; Ammassari-Teule, M. Progressive cognitive decline in a transgenic mouse model of Alzheimer's disease overexpressing mutant *hAPPswe*. *Genes Brain Behav.,* 2006, 5, 249-56.

Middei, S; Geracitano, R; Caprioli, A; Mercuri, N; Ammassari-Teule, M. Preserved fronto-striatal plasticity and enhanced procedural learning in a transgenic mouse model of Alzheimer's disease overexpressing mutant *hAPPswe*. *Learn. Mem.,* 2004, 11, 447-52.

Minkeviciene, R; Ihalainen, J; Malm, T; Matilainen, O; Keksa-Goldsteine, V; Goldsteins, G; Iivonen, H; Leguit, N; Glennon, J; Koistinaho, J; Banerjee, P; Tanila, H. Age-related decrease in stimulated glutamate release and vesicular glutamate transporters in APP/PS1 transgenic and wild-type mice. *J. Neurochem.,* 2008, 105, 584-94.

Moechars, D; Dewachter, I; Lorent, K; Reversé, D; Baekelandt, V; Naidu, A; Tesseur, I; Spittaels, K; Van Den Haute, C; Checler, F; Godaux, E; Cordell, B; Van Leuven, F. Early phenotypic changes in transgenic mice that overexpress different mutants of amyloid precursor protein in brain. *J. Biol. Chem.,* 1999, 274, 6483-92.

Moran, PM; Higgins, LS; Cordell, B; Moser, PC. Age-related learning deficits in transgenic mice expressing the 751-amino acid isoform of human β-amyloid protein precursor protein. *Proc. Natl. Acad. Sci. USA,* 1995, 92, 5341-5.

Morgan, D. Learning and memory deficits in APP transgenic mouse models of amyloid deposition. *Neurochem. Res.,* 2003, 28, 1029-34.

Morris, RGM; Garrud, P; Rawlins, JNP; O'Keefe, J. Place navigation impaired in rats with hippocampal lesions. *Nature,* 1982, 292, 681-3.

Mucke, L; Masliah, E; Johnson, WB; Ruppe, MD; Alford, M; Rockenstein, EM; Forss-Petter, S; Pietropaolo, M; Mallory, M; Abraham, CR. Synaptotrophic effects of human amyloid beta protein precursors in the cortex of transgenic mice. *Brain Res.,* 1994, 666, 151-67.

Mucke, L; Masliah, E; Yu, G-Q; Mallory, M; Rockenstein, EM; Tatsuno, G; Hu, K; Kholodenko, D; Johnson-Wood, K; McConlogue, L. High-level neuronal expression of $A\beta_{1-42}$ in wild-type human amyloid protein precursor transgenic mice: synaptotoxicity without plaque formation. *J. Neurosci.,* 2000, 20, 4050-8.

Müller, U; Cristina, N; Li, Z-W; Wolfer, DP; Lipp, H-P; Rülicke, T; Brandner, S; Aguzzi, A; Weissmann, C. Behavioral and anatomical deficits in mice homozygous for a modified β-amyloid precursor protein gene. *Cell,* 1994, 79, 755-65.

Nalbantoglu, J; Tirado-Santiago, G; Lahsaïni, A; Poirier, J; Goncalves, O; Verge, G; Momoli, F; Welner, SA; Massicotte, G; Julien, J-P; Shapiro, ML. Impaired learning and LTP in mice expressing the carboxy terminus of the Alzheimer amyloid precursor protein. *Nature,* 1997, 387, 500-5.

Nelson, RL; Guo, Z; Halagappa, VM; Pearson, M; Gray, AJ; Matsuoka, Y; Brown, M; Martin, B; Iyun, T; Maudsley, S; Clark, RF; Mattson, MP. Prophylactic treatment with paroxetine ameliorates behavioral deficits and retards the development of amyloid and tau pathologies in 3xTgAD mice. *Exp. Neurol.,* 2007, 205, 166-76.

Nicolakakis, N; Aboulkassim, T; Ongali, B; Lecrux, C; Fernandes, P; Rosa-Neto, P; Tong, XK; Hamel, E. Complete rescue of cerebrovascular function in aged Alzheimer's disease transgenic mice by antioxidants and pioglitazone, a peroxisome proliferator-activated receptor gamma agonist. *J. Neurosci.,* 2008, 28, 9287-96.

Nilsson, LNG; Arendash, GW; Leighty, RE; Costa, DA; Low, MA; Garcia, MF; Cracciolo, JR; Rojiani, A; Wu, X; Bales, KR; Paul, SM; Potter, H. Cognitive impairment in PDAPP mice depends on ApoE and ACT-catalyzed amyloid formation. *Neurobiol. Aging*, 2004, 25, 1153-67.

Nishida, Y; Yokota, T; Takahashi, T; Uchihara, T; Jishage, K; Mizusawa, H. Deletion of vitamin E enhances phenotype of Alzheimer disease model mouse. *Biochem. Biophys. Res. Comm.,* 2006, 350, 530-6.

Oakley, H; Cole, S.L; Logan, S; Maus, E; Shao, P; Craft, J; Guillozet-Bongaarts, A; Ohno, M; Disterhoft, J; Van Eldik, L; Berry, R; Vassar, R. Intraneuronal β-amyloid aggregates, neurodegeneration, and neuron loss in transgenic mice with five familial Alzheimer's disease mutations: potential factors in amyloid plaque formation. *J. Neurosci.,* 2006, 26, 10129-40.

Oddo, S; Caccamo, A; Shepherd, JD; Murphy, MP; Golde, TE; Kayed, R; Metherate, R; Mattson, MP; Akbari, Y; LaFerla, FM. Triple-transgenic model of Alzheimer's disease with plaques and tangles: intracellular Aβ and synaptic dysfunction. *Neuron,* 2003, 39, 409-21.

Ognibene, E; Middei, S; Daniele, S; Adriani, W; Ghirardi, O; Caprioli, A; Laviola, G. Aspects of spatial memory and behavioral disinhibition in Tg2576 transgenic mice as a model of Alzheimer's disease. *Behav. Brain Res.,* 2005, 156, 225-32.

Ohno, M; Chang, L; Tseng, W; Oakley, H; Citron, M; Klein, WL; Vassar, R; Disterhoft, JF. Temporal memory deficits in Alzheimer's mouse models: rescue by genetic deletion of BACE1. *Eur. J. Neurosci.*, 2006, 23, 251-60.

Olcese, JM; Cao, C; Mori, T; Mamcarz, MB; Maxwell, A; Runfeldt, MJ; Wang, L; Zhang, C; Lin, X; Zhang, G; Arendash, GW. Protection against cognitive deficits and markers of neurodegeneration by long-term oral administration of melatonin in a transgenic model of Alzheimer disease. *J. Pineal Res.,* 2009, 47, 82-96.

O'Leary, TP; Brown, RE. Visuo-spatial learning and memory deficits on the Barnes maze in the 16-month-old APPswe/PS1dE9 mouse model of Alzheimer's disease. *Behav. Brain Res.,* 2009, 201, 120-7.

O'Reilly, JA; Lynch, M. Rosiglitazone improves spatial memory and decreases insoluble $A\beta_{1,42}$ in APP/PS1 mice. *J. Neuroimmune Pharmacol.*, 2012, 7, 140-4.

Oster-Granite, ML; McPhie, DL; Greenan, J; Neve, RL. Age-dependent neuronal and synaptic degeneration in mice transgenic for the C terminus of the amyloid precursor protein. *J. Neurosci.,* 1996, 16, 6732-41.

Palop, JJ; Jones, B; Kekonius, L; Chin, J; Yu, GQ; Raber, J; Masliah, E; Mucke, L. Neuronal depletion of calcium-dependent proteins in the dentate gyrus is tightly linked to Alzheimer's disease-related cognitive deficits. *Proc. Natl. Acad. Sci. USA,* 2003, 100, 9572-7.

Park, JH; Widi, GA; Gimbel, DA; Harel, NY; Lee, DHS; Strittmayer, SM. Subcutaneous Nogo receptor removes brain amyloid-β and improves spatial memory in Alzheimer's transgenic mice. *J. Neurosci.,* 2006, 26, 13279-86.

Peng, J; Liang, G; Inan, S; Wu, Z; Joseph, DJ; Meng, Q; Peng, Y; Eckenhoff, MF; Wei, H. Dantrolene ameliorates cognitive decline and neuropathology in Alzheimer triple transgenic mice. *Neurosci. Lett.,* 2012, 516, 274-9.

Perez-Cruz, C; Nolte, MW; van Gaalen, MM; Rustay, NR; Termont, A; Tanghe, A; Kirchhoff, F; Ebert, U. Reduced spine density in specific regions of CA1 pyramidal neurons in two transgenic mouse models of Alzheimer's disease. *J. Neurosci.,* 2011, 31, 3926-34.

Phinney, AL; Horne, P; Yang, J; Janus, C; Bergeron, C; Westaway, D. Mouse models of Alzheimer's disease: the long and filamentous road. *Neurol. Res.,* 2003, 25, 590-600.

Pietropaolo, S; Feldon, J; Yee, BK. Age-dependent phenotypic characteristics of a triple transgenic mouse model of Alzheimer's disease. *Behav. Neurosci.,* 2008, 122, 733-47.

Pietropaolo, S; Sun, Y; Li, R; Brana, C; Feldon, J; Yee, BK. Limited impact of social isolation on Alzheimer-like symptoms in a triple transgenic model. *Behav. Neurosci.,* 2009, 123, 181-95.

Pillay, NS; Kellaway, LA; Kotwal, GJ. Early detection of memory deficits and memory improvement with vaccinia virus complement control protein in an Alzheimer's disease model. *Behav. Brain Res.,* 2008, 192, 173-7.

Platt, B; Drever, B; Koss, D; Stoppelkamp, S; Jyoti, A; Plano, A; Utan, A; Merrick, G; Ryan, D; Melis, V; Wan, H; Mingarelli, M; Porcu, E; Scrocchi, L; Welch, A; Riedel, G. Abnormal cognition, sleep, EEG and brain metabolism in a novel knock-in Alzheimer mouse, PLB1. *PLoS One,* 2011, 6(11), e27068.

Poirier, R; Wolfer, DP; Welzl, H; Tracy, J; Galsworthy, MJ; Nitsch, RM; Mohajeri, MH. Neuronal neprilysin overexpression is associated with attenuation of Abeta-related spatial memory deficit. *Neurobiol. Dis.,* 2006, 24, 475-83.

Pompl, PN; Mullan, MJ; Bjugstad, K; Arendash, GW. Adaptation of the circular platform spatial memory task for mice: use in detecting cognitive impairment in the APP_{SW} transgenic mouse model for Alzheimer's disease. *J. Neurosci. Methods,* 1999, 87, 87-95.

Puoliväli, J; Wang, J; Heikkinen, T; Heikkilä, M; Tapiola, T; van Groen, T; Tanila, H. Hippocampal Aβ42 levels correlate with spatial memory deficit in APP and PS1 double transgenic mice. *Neurobiol. Dis.,* 2002, 9, 339-47.

Qin, C; Zhu, H; Zhang, B; Chang, Y; Yin H; Liu, Y. Patho-morphological and immunohistochemical findings in brains of APP transgenic model. *Acta. Lab. Sci. Sinica,* 2000, 8, 213–7.

Quinn, JF; Bussiere, JR; Hammond RS; Montine, TJ; Henson, E; Jones, RE; Stackman, RW Jr. Chronic dietary alpha-lipoic acid reduces deficits in hippocampal memory of aged Tg2576 mice. *Neurobiol. Aging*, 2007, 28, 213-25

Quon, D; Wang, Y; Catalano, R; Scardina, JM; Murakami, K; Cordell, B. Formation of β-amyloid protein deposits in brains of transgenic mice. *Nature,* 1991, 352, 239-41.

Radde, R; Bolmont, T; Kaeser, SA; Coomaraswamy, J; Lindau, D; Stoltze, L; Calhoun, ME; Jäggi, F; Wolburg, H; Gengler, S; Haass, C; Ghetti, B; Czech, C; Hölscher, C; Mathews, PM; Jucker, M. Aβ42-driven cerebral amyloidosis in transgenic mice reveals early and robust pathology. *EMBO Rep.,* 2006, 7, 940-6.

Reed, MN; Liu, P; Kotilinek, LA; Ashe, KH. Effect size of reference memory deficits in the Morris water maze in Tg2576 mice. *Behav. Brain Res.,* 2010, 212, 115-20.

Reiserer, RS; Harrison, FE; Syverud, DC; McDonald, MP. Impaired spatial learning in the APP_{Swe}+PSEN1ΔE9 bigenic mouse model of Alzheimer's disease. *Genes Brain Behav.,* 2007, 6, 54-65.

Ribé, EM; Pérez, M; Puig, B; Gich, I; Lim, F; Cuadrado, M; Sesma, T; Catena, S; Sanchez, B; Nieto, M; Gomez-Ramos, P; Moran, MA; Cabodevilla, F; Samaranch, L; Ortiz, L; Pérez, A; Ferrer, I; Avila, J; Gomez-Isla, T. Accelerated amyloid deposition, neurofibrillary degeneration and neuronal loss in double mutant APP/tau transgenic mice. *Neurobiol. Dis.,* 2005, 20, 814-22.

Ribes, D; Colomina, MT; Vicens, P; Domingo, JL. Effects of oral aluminum exposure on behavior and neurogenesis in a transgenic mouse model of Alzheimer's disease. *Exp. Neurol.,* 2008, 214, 293-300.

Richardson, JC; Kendal, CE; Anderson, R; Priest, F; Gower, E; Soden, P; Gray, R; Topps, S; Howlett, DR; Lavender, D; Clarke, NJ; Barnes, JC; Haworth, R; Stewart, MG; Rupniak, HTR. Ultrastructural and behavioural changes precede amyloid deposition in a transgenic model of Alzheimer's disease. *Neuroscience,* 2003, 122, 213-28.

Ricobaraza, A; Cuadrado-Tejedor, M; Pérez-Mediavilla, A; Frechilla, D; Del Rio, J; Garcia-Osta, A. Phenylbutyrate ameliorates cognitive deficit and reduces tau pathology in an Alzheimer's disease mouse model. *Neuropsychopharmacology,* 2009, 34, 1721-32.

Sadowski, M; Pankiewicz, J; Scholtzova, H; Ji, Y; Quartermain, D; Jensen, CH; Duff, K; Nixon, RA; Gruen, RJ; Wisniewski, T. Amyloid-β deposition is associated with decreased hippocampal glucose metabolism and spatial memory impairment in APP/PS1 mice. *J. Neuropathol. Exp. Neurol.,* 2004, 63, 418-28.

Savonenko, AV; Xu, GM; Price, DL; Borchelt, DR; Markowska, AL. Normal cognitive behavior in two distinct congenic lines of transgenic mice hyperexpressing APP_{SWE}. *Neurobiol. Dis.,* 2003, 12, 194-211.

Savonenko, A; Xu, GM; Melnikova, T; Morton, JL; Gonzales, V; Wong, MP; Price, DL; Tang, F; Markowska, AL; Borchelt, DR. Episodic-like memory deficits in the APPswe/PS1dE9 mouse model of Alzheimer's disease: relationships to β-amyloid deposition and neurotransmitter abnormalities. *Neurobiol. Dis.*, 2005, 18, 602-17.

Scearce-Levie, K. Monitoring spatial learning and memory in Alzheimer's disease mouse models using the Morris water maze. In: ED Roberson (ed), *Alzheimer's disease and frontotemporal dementia* (Methods in Molecular Biology, vol 670). Berlin: Springer, pp 191-205, 2011.

Scholtzova, H; Kascsak, RJ; Bates, KA; Boutajangout, A; Kerr, D.J; Meeker, H.C; Mehta, PD; Spinner, DS; Wisniewski, T. Induction of toll-like receptor 9 signaling as a method for ameliorating Alzheimer's disease-related pathology. *J. Neurosci.,* 2009, 29, 1846-54.

Seabrook, GR; Rosahl, TW. Transgenic animals relevant to Alzheimer's disease. *Neuropharmacology*, 1999, 8, 1-17.

Sigurdsson, EM; Knudsen, E; Asuni, A; Fitzer-Attas, C; Sage, D; Quartermain, D; Goni, F; Frangione, B; Wisniewski, T. An attenuated immune response is sufficient to enhance cognition in an Alzheimer's disease mouse model immunized with amyloid-β derivatives. *J. Neurosci.,* 2004, 24, 6277-82.

Simón, AM; Schiapparelli, L; Salazar-Colocho, P; Cuadrado-Tejedor, M; Escribano, L; López de Maturana, R; Del Río, J; Pérez-Mediavilla, A; Frechilla, D. Overexpression of wild-type human APP in mice causes cognitive deficits and pathological features unrelated to Aβ levels. *Neurobiol. Dis.,* 2009, 33, 369-78.

Spencer, B; Marr, RA; Rockenstein, E; Crews, L; Adame, A; Potkar, R; Patrick, C; Gage, FH; Verma, IM; Masliah, E. Long-term neprilysin transfer is associated with reduced levels of intracellular Abeta and behavioral improvement in APP transgenic mice. *BMC Neurosci.,* 2008, 9, 109.

Stewart, S; Cacucci, F; Lever, C. Which memory task for my mouse? A systematic review of spatial memory performance in the Tg2576 Alzheimer's mouse model. *J. Alzheimers Dis.,* 2011, 26, 105-26.

Stover, KR; Brown, RE. Age-related changes in visual acuity, learning and memory in the APPswe/PS1dE9 mouse model of Alzheimer's disease. *Behav. Brain Res.,* 2012, 231, 75-85.

Sturchler-Pierrat, C; Abramowski, D; Duke, M; Wiederhold, K-H; Mislt, C; Rothacher, S; Ledermann, B; Bürki, K; Frey, P; Paganetti, PA; Waridel, C; Calhoun, ME; Jucker, M; Probst, A; Staufenbiel, M; Sommer, B. Two precursor protein transgenic mouse models with Alzheimer disease-like pathology. *Proc. Natl. Acad. Sci. USA,* 1997, 94, 13287-92.

Su, D; Zhao, Y; Xu, H; Wang, B; Chen, X; Chen, J; Wang, X. Isoflurane exposure during mid-adulthood attenuates age-related spatial memory impairment in APP/PS1 transgenic mice. *PLoS One,* 2012, 7(11), e50172.

Sykova, E; Vorisek, I; Antonova, T; Mazel, T; Meyer-Luehmann, M; Jucker, M; Hajek, M; Or, M; Bures, J. Changes in extracellular space size and geometry in APP23 transgenic mice: A model of Alzheimer's disease. *Proc. Natl. Acad. Sci .USA,* 2005, 102, 479-84.

Tampellini, D; Capetillo-Zarate, E; Dumont, M; Huang, Z; Yu, F; Lin, MT; Gouras, GK. Effects of synaptic modulation on β-amyloid, synaptophysin, and memory performance in Alzheimer's disease transgenic mice. *J. Neurosci.,* 2010, 30, 14299-304.

Tanila, H. Wading pools, fading memories-place navigation in transgenic mouse models of Alzheimer's disease. *Front. Aging Neurosci.,* 2012, 4, 1-7.

Tiraboschi, P; Hansen, LA; Thal, LJ; Corey-Bloom, J. The importance of neuritic plaques and tangles to the development and evolution of AD. *Neurology,* 2004a, 62, 1984-9.

Tiraboschi, P; Sabbagh, MN; Hansen, LA; Salmon, DP; Merdes, A; Gamst, A; Masliah, E; Alford, M; Thal, LJ; Corey-Bloom, J. Alzheimer disease without neocortical neurofibrillary tangles: "a second look". *Neurology,* 2004b, 62, 1141-7.

Tomiyama, T; Matsuyama, S; Hiroyuki, I; Umeda, T; Takuma, H; Ohnishi, K; Ishibashi, K; Teraoka, R; Sakama, N; Yamashita, T; Nishitsuji, K; Ito, K; Shimada, H; Lambert, MP; Klein, WL; Mori, H. A mouse model of amyloid *β* oligomers: Their contribution to synaptic alteration, abnormal tau phosphorylation, glial activation, and neuronal loss in vivo. *J. Neurosci*., 2010, 30, 4845-56.

Tremml, P; Lipp, H-P; Müller, U; Ricceri, L; Wolfer, DP. Neurobehavioral development, adult openfield exploration and swimming navigation learning in mice with a modified β-amyloid-precursor protein gene. *Behav. Brain Res.,* 1998, 95, 65-76.

Trinchese, F; Liu, S; Battaglia, F; Walter, S; Mathews, PM; Arancio, O. Progressive age-related development of Alzheimer-like pathology in APP/PS1 mice. *Ann. Neurol.,* 2004, 55, 801-14.

Umeda, T; Tomiyama, T; Kitajima, E; Idomoto, T; Nomura, S; Lambert, MP; Klein, WL; Mori, H. Hypercholesterolemia accelerates intraneuronal accumulation of Aβ oligomers resulting in memory impairment in Alzheimer's disease model mice. *Life Sci.,* 2012, 91, 1169-76.

Van Dam, D; D'Hooge, R; Staufenbiel, M; Van Ginneken, C; Van Meir, F; De Deyn, PP. Age-dependent cognitive decline in the APP23 model precedes amyloid deposition. *Eur. J. Neurosci.,* 2003, 17, 388-96.

Van Dijck, A; Van Dam, D; De Deyn, PP. NAP has no effect on spatial memory after short-term treatment in advanced stage Alzheimer's disease mouse model. *Peptides,* 2009, 30, 2480-2.

van Leuven, FF. Single and multiple transgenic mice as models for Alzheimer's disease. *Prog. Neurobiol.,* 2000, 61, 305-12.

Vloeberghs, E; Van Dam, D; D'Hooge, R; Staufenbiel, M; De Deyn, PP. APP23 mice display working memory impairment in the plus-shaped water maze. *Neurosci. Lett.,* 2006, 407, 6-10.

Walker, JM; Fowler, SW; Miller, DK; Sun, AY; Weisman, GA; Wood, WG; Sun, GY; Simonyi, A; Schachtman, TR. Spatial learning and memory impairment and increased locomotion in a transgenic amyloid precursor protein mouse model of Alzheimer's disease. *Behav. Brain Res.,* 2011, 222, 169-75.

Wang, C-M; Devries, S; Camboni, M; Glass, M; Martin, PT. Immunization with the SDPM1 peptide lowers amyloid plaque burden and improves cognitive function in the APPswePSEN1(A246E) transgenic mouse model of Alzheimer's disease. *Neurobiol. Dis.,* 2010, 39, 409-22.

Westerman, MA; Cooper-Blacketer, D; Mariash, A; Kotilinek, L; Kawarabayashi, T; Younkin, LH; Carlson, GA; Younkin, SG; Ashe, KH. The relationship between Aβ and memory in the Tg2576 mouse model of Alzheimer's disease. *J. Neurosci.,* 2002, 22, 1858-67.

Wilcock, DM; Rojiani, A; Rosenthal, A; Subbarao, S; Freeman, MJ; Gordon, MN; Morgan, D. Passive immunotherapy against Aβ in aged APP-transgenic mice reverses cognitive deficits and depletes parenchymal amyloid deposits in spite of increased vascular amyloid and microhemorrhage. *J. Neuroinflamm.*, 2004, 1, 24.

Wilcock, DM; Lewis, MR; Van Nostrand, WE; Davis, J; Previti, ML; Gharkholonarehe, N; Vitek, MP; Colton, CA. Progression of amyloid pathology to Alzheimer's disease pathology in an amyloid precursor protein transgenic mouse model by removal of nitric oxide synthase 2. *J. Neurosci.,* 2008, 28, 1537-45.

Wolfer, DP; Müller, U; Stagliar, M; Lipp, H-P. Assessing the effects of the 129/Sv genetic background on swimming navigation learning in transgenic mutants: a study using mice with a modified *β-amyloid-precursor protein* gene. *Brain Res.*, 1997, 771, 1-13.

Yamada, M; Ihara, M; Okamoto, Y; Maki, T; Washida, K; Kitamura, A; Hase, Y; Ito, H; Takao, K; Miyakawa, T; Kalaria, RN; Tomimoto, H; Takahashi, R. The influence of chronic cerebral hypoperfusion on cognitive function and amyloid β metabolism in APP overexpressing mice. *PLoS One,* 2011, 6, e16567.

Yao, J; Du, H; Yan, S.; Wang, C; Lue, LF; Guo, L; Chen, D; Stern, DM; Gunn Moore, FJ; Chen, JXI; Arancio, O; Yan, SS. Inhibition of amyloid-β (Aβ) peptide-binding alcohol dehydrogenase-Aβ interaction reduces Aβ accumulation and improves mitochondrial function in a mouse model of Alzheimer's disease. *J. Neurosci.,* 2011, 31, 2313-20.

Yamaguchi, F; Richards, S-J; Beyruther, K; Salbaum, M; Carlson, GA; Dunnett, SB. Transgenic mice for the amyloid precursor protein 695 isoform have impaired spatial memory. *NeuroReport,* 1991, 2, 781-4.

Yoshiike, Y; Kimura, T; Yamashita, S; Furudate, H; Mizoroki, T; Murayama, M; Takashima A. $GABA_A$ receptor-mediated acceleration of aging-associated memory decline in APP/PS1 mice and its pharmacological treatment by picrotoxin. *PLoS One*, 2008, 3(8), e3029.

Zhang, L; Xing, Y; Ye, CF; Ai, HX; Wei, HF; Li, L. Learning-memory deficit with aging in APP transgenic mice of Alzheimer's disease and intervention by using tetrahydroxystilbene glucoside. *Behav. Brain Res.,* 2006, 173, 246-54.

Zhang, W; Bai, M; Xi, Y; Hao, J; Liu, L; Mao, N; Su, C; Miao, J; Li, Z. Early memory deficits precede plaque deposition in APPswe/PS1dE9 mice: Involvement of oxidative stress and cholinergic dysfunction. *Free Radic. Biol. Med.,* 2012, 52, 1443-52.

Zhang, X; Li, L; Zhang, X; Xie, W; Li, L; Yang, D; Heng, X; Du, Y; Doody, R.S; Le, W. Prenatal hypoxia may aggravate the cognitive impairment and Alzheimer's disease neuropathology in APP(Swe)/PS1(A246E) transgenic mice. *Neurobiol. Aging,* 2013, 34, 663-78.

Zhuo, JM; Prescott, SL; Murray, ME; Zhang, HY; Baxter, MG; Nicolle, MM. Early discrimination reversal learning impairment and preserved spatial learning in a longitudinal study of Tg2576 APPsw mice. *Neurobiol. Aging,* 2007, 28, 1248-57.

In: Spatial Orientation in Humans and Animals
Editor: Robert Lalonde
ISBN: 978-1-63463-001-6
© 2014 Nova Science Publishers, Inc.

Chapter 6

SPATIAL DEFICITS IN PATIENTS WITH PARKINSON'S AND HUNTINGTON'S DISEASES

Robert Lalonde*
Université de Rouen, Dépt. Psychologie,
Mont-Saint-Aignan Cedex France

1. PARKINSON'S DISEASE

1.1. Neuropathology and Behavioral Symptoms

Parkinson's disease is a neurodegenerative condition mainly affecting the basal ganglia (Marsden, 1989; Miller and DeLong, 1988). Cell loss occurs in substantia nigra pars compacta and ventral tegmental area, which depletes dopamine concentrations in dorsal and ventral striatum, as well as nucleus accumbens and prefrontal cortex (chapter 3). Dopamine concentrations are also depleted in retina.

The basal ganglia neuropathology results in anomalies of neocortico-striato-pallido-thalamo-neocortical pathways underlying cognitive and motor functions (Hoshi, 2013; Miller and DeLong, 1988).

The cardinal symptoms in Parkinson's disease include akinesia (slowness in initiating movements), bradykinesia (slowness during movements), muscular rigidity, and resting tremor (Miller and DeLong, 1988; Marsden, 1989; Weiner, 1989), treatable with oral administration of the immediate metabolic precursor of dopamine, l-dopa, dopamine receptor agonists, or dopamine reuptake blockers (Blanchet, 1999; Kilpatrick and Tilbrook, 2002). Although a non-dementing disorder as such, Parkinson's disease has a higher probability of leading to Alzheimer and other forms of dementia.

In a sample of over 220,000 Norwegian subjects with Parkinson's disease, the frequency of dementia was 28% (Aarsland et al., 1996).

* Email: robert.lalonde@univ-rouen.fr

1.2. Genetics

Although Parkinson's disease usually manifests itself in the sporadic form, it may also be transmitted through heredity in autosomal dominant or recessive forms (Lee and Liu, 2008). In the autosomal dominant form, Parkinson's disease may be caused by missense A30P and A53T mutations of the *SNCA* gene located on chromosome 4 and encoding α-synuclein, also known as PARK1 (Kruger et al., 1998; Polymeropoulos et al., 1997), the main constituent of the Lewy body, normally localized at synapses (Kahle et al., 2001) and a histologic marker of patients with Parkinson's disease or Lewy body dementia (Spillantini et al., 1997). In the autosomal recessive form, hereditary Parkinson's disease may be caused by a loss in function of *PARK2* located on chromosome 6 and encoding parkin (Kitada et al., 1998), as well as *DJ1* located on chromosome 1 (Bonifati et al., 2003). Polymorphisms of *PARK2* may be responsible for some forms of sporadic Parkinson's disease (Wang et al., 1999). Parkin deletion causes accumulation of α-synuclein, because it promotes ubiquination of this protein prior to degradation by the proteosome (Shimura et al., 2001).

1.3. Visual Acuity and Spatial Attention

Some basic visual functions are deficient in patients with Parkinson's disease (Bodis-Wollner, 1990), namely contrast sensitivity (Bodis-Wollner, et al., 1987; Diederich et al. 2002; Regan and Maxner, 1987), flicker sensitivity (Bodis-Wollner, et al., 1987), and color discrimination (Diederich et al. 2002). Bodis-Wollner (1990) proposed that these deficiencies are caused by retinal depletions of dopamine. To investigate the relation between sensory and dopaminergic deficiencies, Shin et al. (2005) used a grating orientation task in patients before and after long-term antiparkinson therapy. The grating orientation threshold was higher in patients than that of controls but improved after dopamine therapy. Sampaio et al. (2011) separated sensory from attention-related patterns of performance by measuring contrast sensitivity with distinct types of cues. Subjects with Parkinson's disease were impaired in their psychophysical performance when valid attentional cues were given, which indicate a disturbance in spatial mechanisms of covert attention. Thus, both sensory and selective attention appears undermined in Parkinson's disease.

1.4. Line Orientation and Bisection

Tsai et al. (1994) reported that patients with Parkinson's disease were impaired relative to controls in the Benton line orientation test. However, this result was not reproduced by others (Levin et al., 1989, 1991; Stirk and Foreman, 2005). Whether the deficit occurs seems to depend on age, disease duration, and the presence of dementia (Raskin et al., 1990). Another posterior-type task, line bisection, has been reported to be impaired in patients with Parkinson's disease (Laudate et al., 2013; Lee et al., 2001; Miller et al., 2013). These data prompted Miller et al. (2013) to discuss about anterior versus posterior subtypes in this disease. The disturbance probably reflects an attentional bias measurable by eye movement patterns and unequal distribution of dopamine concentrations in each hemisphere (Laudate et al., 2013).

1.5. Visuoconstructive Abilities

As observed in patients with lesions to frontal or parietal association cortex (chapter 1), patients with Parkinson's disease are deficient in several visuoconstructive tasks (Table 1). Relative to healthy controls, patients with Parkinson's disease were impaired in block design (Hietanen and Teräväinen, 1988; Levin et al., 1991) and object assembly (Morris et al., 1988) subtests of the Wechsler intelligence scale. Patients with Parkinson's disease were also impaired in copying simple drawings such as a house, a cube, a star with or without landmarks providing step-by-step organization (Iwasaki et al., 1989), a parietal-like function attributable to disrupted pathways between this region and the basal ganglia (chapter 3).

Table 1. Spatial deficits in patients with Parkinson's disease

Function	Tests	Reference
visuoconstructive skills	block design	Hietanen and Teräväinen, 1988; Levin et al., 1991
visuoconstructive skills	object assembly	Morris et al., 1988
visuoconstructive skills	simple drawings	Iwasaki et al., 1989
visuoconstructive skills	Rey complex figure	Mohr et al., 1990a,b; Stepkina et al., 2010
spatial reasoning	Raven standard or colored progressive matrices	Dubois et al., 1990; Huber et al., 1986
map reading	Money's road-map	Mohr et al., 1990a; Taylor et al., 1986
working memory	Corsi block-tapping	Yágüez et al., 2006
working memory	delayed response	Freedman and Oscar-Berman, 1989b; Partiot et al., 1996; Sahakian et al., 1988
working memory	blue token	Owen et al., 1992, 1997
working memory	recognition memory	Owen et al., 1993
working memory	box location	Fournet et al., 1996
long-term memory	figure reproduction Wechsler memory scale	Dubois et al., 1990; Hietanen and Teräväinen, 1988; Mohr et al., 1990a
long-term memory	recall of Rey or Benton figures	Levin et al., 1989; Mohr et al., 1990a,b
long-term memory	Grober-Buschke selective reminding	Pillon et al., 1996
long-term memory	picture location	Pillon et al., 1996, 1997
long-term memory	conditional associative learning	Pillon et al., 1998; Postle et al., 1997
maze learning	covered mazes	Wallesch et al., 1990
maze learning	route-walking	Bowen et al., 1972

The landmarks score was correlated with the bradykinesia score, a result that implicates dopamine for both measures. Crucian and Okun (2003) conclude that although often relevant, fronto-basal ganglionic dysfunction accounts only partly for the spatial deficits seen in this disease. As seen with simpler drawings, patients with Parkinson's disease were impaired relative to controls in copying the Rey complex figure (Mohr et al., 1990a,b) and their performance deteriorated over a period of a few years (Stepkina et al., 2010). However, in one study, the patients performed this task normally (Sprengelmeyer et al., 1995). The variability in results may be caused by different levels of disease progression or to the benefits of dopaminergic medication. The Rey complex figure was one of the tests reaching screening and diagnostic validity relative to normative data for early cognitive impairment in Parkinson's disease (Biundo et al., 2013).

1.6. Spatial Reasoning

As described in chapter 1, spatial reasoning may be estimated with Raven standard progressive matrices or a simpler version of the same test, the Raven colored progressive matrices. The performances of Raven standard progressive matrices (Huber et al., 1986) or colored progressive matrices (Dubois et al., 1990) were poorer in patients with Parkinson's disease than those of controls. However, in a separate study, their scores on colored progressive matrices were normal despite impairments in the Wisconsin card sorting test of concept formation (Caltagirone et al., 1989). These results indicate that spatial reasoning may be less vulnerable to this disease, especially in the early stage, than more prefrontally mediated tasks.

1.7. Map Reading

Spatial deficits in patients with Parkinson's disease are evident in map reading tests (chapter 1). In particular, patients with Parkinson's disease were impaired on the Money road-map test of orientation sense (Mohr et al., 1990a; Taylor et al., 1986). On the contrary, parkinsonian subjects were not affected in a semantic memory task requiring the answer to the simple question as to which street the subject lives on (Rodríguez-Ferreiro et al., 2010).

Two other mapping tasks have been examined in the patients, one requiring recall of a layout of 12 places on a map and the other a layout of 12 dots (Giraudo et al., 1997). The patients performed more poorly than controls only the dot-layout task. These results emphasize the sometimes subtle nature of the deficit.

It is probable that the dot-layout task appeared more conceptually abstract than the place task. In three-dimensional testing methods, subjects were led inside a room containing 4 objects, memorized their location, and, after being led to another room, placed icons representing them on a map (Stepánková et al., 1998). In a second test, subjects chose a map showing the correct arrangement of objects from 4 choices, the locations of some objects being changed. There were no intergroup differences on either version, indicating once again that spatial memory of parkinsonian subjects appears normal on maps presented under conditions approaching real-life settings.

1.8. Spatial Short-Term or Working Memory

On one hand, patients with Parkinson's disease were impaired in a computerized version of the Corsi block-tapping test of spatial short-term memory (Yágüez et al., 2006). On the other, the patients performed normally on standard (Levin et al., 1989) or computerized (Morris et al., 1988) versions of this test. In an adaptation of the Corsi block-tapping task, the performance of medicated but not unmedicated patients with early-stage Parkinson's disease was deficient (Stoffers et al., 2003). The medicated subjects were probably more vulnerable because of their being further along disease progression in a manner uncompensated for by dopaminergic therapy. Thus, the negative results may be due to the analysis of less advanced cases. In a pattern span paradigm, the patients were impaired at stimulus encoding, storage, or response programming stages in comparison with control subjects (Le Bras et al., 1999).

Left-right discrimination learning and spatial reversal, two tests used in animal research (chapters 1 and 2), were analyzed in demented patients with Parkinson's disease (Freedman and Oscar-Berman, 1989a). Spatial reversal learning but not left-right discrimination was impaired in patients relative to controls, which implies a selective deficit in mental flexibility mediated by the prefrontal cortex. Two other tasks reminiscent of animal research have been conducted in demented patients with Parkinson's disease: delayed response and left-right alternation, both dependent on working memory and the prefrontal cortex (chapters 1 and 2). The patients were impaired relative to controls for delayed response but not left-right alternation (Freedman and Oscar-Berman, 1989b). Likewise, delayed alternation was normal in patients in the early stages of Parkinson's disease (Canavan et al., 1990). In contrast, a delayed response deficit in patients with Parkinson's disease was confirmed in another study (Partiot et al., 1996). Delayed alternation may be more liable to compensatory strategies than delayed response. In a more complex version of delayed response, subjects were required to memorize locations of abstract visual stimuli varying from 1 to 8 (Sahakian et al., 1988). Unlike controls, medicated and unmedicated patients with Parkinson's disease failed the test with as few as 6 items. The vulnerability of patients with Parkinson's disease to prefrontally mediated functions has been found in many other tests, attributable to prefrontal or striatal depletion of dopamine concentrations, causing dysfunctional neocortical-basal ganglia circuitries (Robbins et al., 1994; Taylor et al., 1986).

In a spatial working memory test comprising a search through boxes to collect blue tokens without returning to the same box twice, a test dependent on the prefrontal cortex (chapter 1), medicated patients with mild or severe clinical symptoms were impaired but not unmedicated patients (Owen et al., 1992, 1997). Once again, the selective deficit in medicated patients is probably due to a more advanced disease progression. The similar deficits found in patients with prefrontal and those with severe basal ganglia damage implicate dysfunctional neocortico-striato-pallido-thalamo-neocortical circuitries as the underlying cause. In another study (Morris et al., 1988), parkinsonian patients were not affected in the blue token task, possibly because disease progression was not sufficiently advanced.

In a similar fashion to the blue token task, medicated patients with severe but not milder symptoms or those without medication were deficient in the spatial recognition memory task (Owen et al., 1993). Likewise, patients with Parkinson's disease were deficient in a spatial working memory task requiring the correct location of boxes in a 5 x 5 matrix (Fournet et al., 1996). In spatial working memory tasks based on reaction times (RTs), patients with Parkinson's disease were slower to respond (Bradley et al., 1989; Brown and Marsden, 1986).

In addition to hand movements, working memory processes may be assessed by eye movements. When subjects were presented with a sequence of four targets which had to be recalled by moving their eyes to fixate the four locations in the correct order, patients with Parkinson's disease had more errors than controls in sequence recall (Hodgson et al., 1999). Spatial working memory deficits are not necessarily due to maintenance in memory. In one study, the patients were impaired only when the to-be-remembered stimulus had a similar location to the probe, but was equivalent across delay intervals (Possin et al., 2008).

Working memory can be analyzed in dual-task paradigms to probe the central executive relative to its two subcomponent parts, the verbal articulatory and the spatial sketch-pad (chapter 1). In one study of this kind, parkinsonian subjects performed the Corsi block-tapping task concurrently with those designed to load on the spatial sketch-pad or the central executive (Kemps et al., 2005). The subjects performed more poorly than controls on both concurrent tasks, which implicates a reduction in cognitive resources of both the spatial sketch-pad and the central executive. When verbal, spatial, and double span tasks were given with varying retention intervals, medicated patients with Parkinson's disease had lower spans relative to controls in all three tasks (Fournet et al., 2000). But when the patients were studied either on or off medication, those off medication exhibited lower spans only in the double task, results that specifically implicate a disturbance in the central executive as a result of dopamine depletion. Likewise, patients with Parkinson's disease had poorer scores in a different spatial working memory task after withdrawal from l-dopa (Lange et al., 1992). Moreover, the subnormal spatial working memory scores of patients with Parkinson's disease improved after receiving l-dopa (Beato et al. 2008). In a review paper, Antal et al. (1998) conclude that spatial working memory is often susceptible to be impaired in Parkinson's disease and depends on dopamine concentrations.

1.9. Long-Term Spatial Memory

Patients with Parkinson's disease have been shown to be deficient in several tasks requiring long-term memory of drawings. For example, the patients were impaired in figure reproduction of the Wechsler memory scale (Dubois et al., 1990; Hietanen and Teräväinen, 1988; Mohr et al., 1990a) as well as recall of the Rey complex figure and Benton figures (Levin et al., 1989; Mohr et al., 1990a,b), along with recall and recognition of a spatial version of the Grober-Buschke selective reminding test (Pillon et al., 1996). Figure reproduction in the Wechsler memory scale did not improve with antiparkinsonian medication, which implies a role for non-dopaminergic mechanisms (Taylor et al., 1987). It must be mentioned that parkinsonians sometimes recall normally Rey and Benton figures (Sprengelmeyer et al., 1995) or figure reproduction in the Wechsler memory scale (Taylor et al., 1986). These variable results are probably due to variable disease durations or to an interaction between disease duration and dopaminergic medications. Likewise, only a severely impaired medicated group was deficient in spatial paired associate learning (Owen et al., 1993).

Long-term spatial memory deficits resulting from Parkinson's disease have been documented in the location of pictures (Pillon et al., 1996), in which even unmedicated patients were affected (Pillon et al., 1997). When patients with mild to moderate Parkinson's disease were investigated in a search task reminiscent of radial-arm maze and Morris water maze paradigms in which participants found and remembered five out of twenty hidden locations, their performance deteriorated when the starting position was moved and proximal cues were deleted, a result that underlines the spatial nature of the deficit. The deficit was related to specific measures of attention. In addition, patients with Parkinson's disease were impaired in a spatial conditional associative learning task (Postle et al., 1997). In particular, both medicated and unmedicated parkinsonian patients were impaired in spatial conditional associative learning (Pillon et al., 1998). The patients were impaired in situations requiring strategic processes and also displayed reduced attentional resources.

Maze learning tasks requiring hand movements are dependent on temporal, parietal, and frontal cortical association cortex (chapter 1). Patients with Parkinson's disease were slower than controls in acquiring a computerized version of covered maze problems (Wallesch et al., 1990). Patients with Parkinson's disease were also impaired in learning routes requiring whole body movements (Bowen et al., 1972) or in orientating to different parts of their own body (Bowen et al., 1976). As reported in the delayed response task described above, these deficits are likely the result of dysfunctional prefrontal-striatal loops secondary to dopamine depletion.

2. Huntington's Disease

2.1. Neuropathology and Behavioral Symptoms

Huntington's disease is a hereditary condition caused by cytosine-adenine-guanine (CAG) expansions of the *HD* gene located on chromosome 4 (Huntington's disease collaborative group, 1993).

The neuropathology affects mainly the basal ganglia, including striatum, globus pallidus, and ventrolateral thalamus (Roos, 1986; Postert et al., 1999). In the dorsal striatum, small spiny striatofugal neurons using gamma-amino butyric acid (GABA) as the neurotransmitter are more affected than large cholinergic interneurons. The neocortical atrophy is most prominent in the prefrontal cortex than the other lobes. Milder or more variable cell losses occur in susbtantia nigra and cerebellar Purkinje cells.

In the early stage, motor symptoms include choreic movements and hyperactivity and cognitive symptoms include anterograde amnesia and a dysexecutive syndrome, eventually leading to dementia (Jacobs et al., 1997). Disease onset is correlated with the number of CAG repeat expansions of the *HD* gene, more copies leading to earlier onset (Andrew et al., 1993; Gusella and MacDonald, 1995).

Neuropsychological anomalies occurred in mutated gene carriers prior to motor signs (Rosenberg et al., 1995) and cognitive scores deteriorated faster with greater CAG repeat expansions (Jason et al., 1997). As seen with Parkinson's disease, the basal ganglia neuropathology causes anomalies of neocortico-striato-pallido-thalamo-neocortical circuitries underlying motor and cognitive functions. Indeed, atrophy at the level of the dorsal striatum based on computerized tomography scans correlated with cognitive deterioration including spatial functions (Bamford et al., 1989). The analysis of spatial orientation has a practical impact in many real-life situations, notably automobile driving, liable to deteriorate in patients with Huntington's disease (Rebok et al., 1995).

Table 2. Spatial deficits in patients with Huntington's disease

Function	Tests	Reference
visuoconstructive skills	block design and object assembly	Bäckman et al., 1997; Boll et al., 1974; Butters et al., 1978; Fedio et al., 1979; Josiassen et al., 1983; Mohr er al, 1991
visuoconstructive skills	Rey complex figure	Bäckman et al., 1997; Gómez-Tortos et al., 1996
visuoconstructive skills	clock drawing	Rouleau et al., 1992
map reading	Money's road-map	Brouwers et al., 1984; Bylsma et al., 1992; Fedio et al., 1979; Mohr et al., 1991
mental rotation	geometric figures	Mohr et al., 1991
working memory	Corsi block-tapping	Yágüez et al., 2006
working memory	recognition memory	Lawrence et al., 2000
working memory	blue tokens	Lawrence et al., 2000
long-term memory	figure reproduction Wechsler memory scale	Tröster et al., 1993
long-term memory	recall of Rey complex figure	Bäckman et al., 1997
long-term memory	stylus maze	Fedio et al., 1979
long-term memory	route-walking	Bylsma et al., 1992
long-term memory	blue velvet arena	Majerová et al., 2012

2.2. Visual Acuity and Spatial Attention

Subtle defects in visual perception have been reported in Huntington's disease, which must be taken into account when evaluating spatial cognition. Relative to nongene carriers, subjects with Huntington's disease or prediagnostic gene carriers had impaired contrast sensitivity for moving gratings (O'Donnell et al., 2008). There were no intergroup differences for contrast sensitivity with the stationary grating or noise thresholds for form and dot motion discriminations. Moreover, there was no intergroup difference in line bisection (Gómez-Tortos et al., 1996). Huntington's disease has also been shown to cause deficits in selective attention for spatial material. Indeed, when patients with Huntington's disease performed a cued RT task with peripheral cues, they were impaired in attentional disengaging from cued locations (Couette et al., 2008).

2.3. Visuoconstructive Abilities

In Huntington's disease, performance IQ is generally more affected than verbal IQ (Taylor et al., 1983). Relative to normal controls, patients with Huntington's disease were impaired in block design and object assembly subtests of the Wechsler intelligence scale (Bäckman et al., 1997; Boll et al., 1974; Butters et al., 1978; Fedio et al., 1979; Josiassen et al., 1983; Mohr et al., 1991).

In addition, patients with Huntington's disease were impaired while copying the Rey complex figure (Bäckman et al., 1997; Gómez-Tortos et al., 1996) or clocks (Rouleau et al., 1992). Drawing of the Rey complex figure was also deficient in a heteregenous group of patients with basal ganglia lesions mainly affecting the medial striatum (caudate nucleus) (Mendez et al., 1989), which indicates a probable role for this brain region in the visuoconstructive troubles underlying Huntington's disease.

2.4. Map Reading

Several experiments have been conducted regarding the impact of Huntington's disease on map reading abilities. Patients with Huntington's disease have often been shown to be impaired in the Money road-map test of orientation sense (Brouwers et al., 1984; Bylsma et al., 1992; Fedio et al., 1979; Mohr et al., 1991). As found with other cognitive tasks, their road-map scores deteriorated over the course of one year (Snowden et al., 2001).

2.5. Mental Rotation

Patients with Huntington's disease were impaired during mental rotation of geometric figures (Mohr et al., 1991). But in another test of this kind, the patients were deficient relative to controls in speed, not accuracy, and thereby not attributable to a speed/accuracy trade-off (Lineweaver et al., 2005).

2.6. Spatial Short-Term and Working Memory

As found in Parkinson's disease (see above), patients with Huntington's disease were impaired relative to controls in a computerized version of the Corsi block-tapping test of spatial short-term memory (Yágüez et al., 2006). They also had more working memory errors during spatial recognition in the location of geometric figures and in searching through boxes to collect blue tokens (Lawrence et al., 2000).

2.7. Spatial Long-Term Memory

Anterograde amnesia is one of the cardinal symptoms of Huntington's disease, notable for both verbal and visuospatial domains. The latter aspect may be evaluated by asking subjects to draw geometric figures from memory. Patients with Huntington's disease were impaired in recall of the figure reproduction subtest of the Wechsler memory scale (Tröster et al., 1993) and the Rey complex figure (Bäckman et al., 1997).

Relative to controls, patients with Huntington's disease were impaired in learning a stylus maze (Fedio et al., 1979). In an analogous manner, a heteregenous group with basal ganglia lesions affecting mainly medial striatum was impaired in the Porteus maze (Mendez et al., 1989). In a route-walking task with or without body turning, the patients were impaired in the former but not the latter task, which indicates an effect on personal as opposed to extrapersonal space (Bylsma et al., 1992). In a group with Huntington's disease evaluated over time, egocentric and allocentric navigation skills were evaluated in the Blue velvet arena, a human analog of the Morris water maze used in animal research (Majerová et al., 2012). Spatial navigation skills deteriorated with increasing motor impairment in Huntington's disease. Unlike the route-walking task described just above, no difference was observed between egocentric and the allocentric skills in this test, both being deficient.

REFERENCES

Aarsland, D; Tandberg, E; Larsen, JP; Cummings, JL. Frequency of dementia in Parkinson disease. *Arch. Neurol.,* 1996, 53, 538-42.

Andrew, SE; Goldberg, YP; Kremer, B; Telenius, H; Theilmann, J; Adam, S; Starr, E; Squitieri, F; Lin, B; Kalchman, MA; Graham, AK; Hayden, MR. The relationship between trinucleotide (CAG) repeat length and clinical features of Huntington's disease. *Nature Genet.,* 1993, 4, 398-403.

Antal, A; Bandini, F; Kéri, S; Bodis-Wollner, I. Visuo-cognitive dysfunctions in Parkinson's disease. *Clin. Neurosci.,* 1998, 5, 147-52.

Bäckman, L; Robins-Wahlin, TB; Lundin, A; Ginovart, N; Farde, L. Cognitive deficits in Huntington's disease are predicted by dopaminergic PET markers and brain volumes. *Brain,* 1997, 120, 2207-17.

Beato, R; Levy, R; Pillon, B; Vidal, C; du Montcel, ST; Deweer,; Bonnet, AM; Houeto, JL; Dubois, B; Cardoso, F. Working memory in Parkinson's disease patients: clinical features and response to levodopa. *Arq. Neuropsiquiatr.,* 2008, 66, 147-51.

Biundo, R; Weis, L; Pilleri, M; Facchini, S; Formento-Dojot, P; Vallelunga, A; Antonini, A. Diagnostic and screening power of neuropsychological testing in detecting mild cognitive impairment in Parkinson's disease. *J. Neural Transm.,* 2013, 120, 627-33.

Blanchet, PJ. Rationale for use of dopamine agonists in Parkinson's disease: review of ergot derivatives. *Can. J. Neurol. Sci.,* 1999, 26, Suppl 2, S21-6.

Bodis-Wollner, I. Visual deficits related to dopamine deficiency in experimental animals and Parkinson's disease patients. *Trends Neurosci.,* 1990, 13, 297-302.

Bodis-Wollner, I; Marx, MS; Mitra, S; Bobak, P; Mylin, L; Yahr, M. Visual dysfunction in Parkinson's disease: loss in spatiotemporal contrast sensitivity. *Brain,* 1987, 110, 1675-98.

Boll, TJ; Heaton, R; Reitan, RM. Neuropsychological and emotional correlates of Huntington's chorea. *J. Nerv. Ment. Dis.,* 1974, 158, 61-9.

Bonifati, V; Rizzu, P; van Baren, MJ; Schaap, O; Breedveld, GJ; Krieger, E; Dekker, MC; Squitieri, F; Ibanez, P; Joosse, M; van Dongen, JW; Vanacore, N; van Swieten, JC; Brice, A; Meco, G; van Duijn, CM; Oostra, BA; Heutink, P. Mutations in the *DJ-1* gene associated with autosomal recessive early-onset parkinsonism. *Science,* 2003, 299, 256-9.

Bowen, FP; Hoehn, MM; Yahr, MD. Parkinsonism, alterations in spatial orientation as determined by a route-walking test. *Neuropsychologia,* 1972, 10, 355-61.

Bowen, FP; Burns, MM; Brady, EM; Yahr, MD. A note on alterations of personal orientation in parkinsonism. *Neuropsychologia,* 1976, 14, 425-9.

Bradley, VA; Welch, JL; Dick, DJ. Visuospatial working memory in Parkinson's disease. *J. Neurol. Neurosurg. Psychiatry,* 1989, 52, 1228-35.

Brouwers, P; Cox, C; Martin, A; Chase, T; Fedio, P. Differential perceptual-spatial impairment in Huntington's and Alzheimer's disease. *Arch. Neurol.,* 1984, 41, 1073-6.

Brown, RG; Marsden, CD. Visuospatial function in Parkinson's disease. *Brain,* 1986, 109, 987-1002.

Butters, N; Sax, D; Montgomery, K; Tarlow, S. Comparison of the neuropsychological deficits associated with early and advanced Huntington's disease. *Arch. Neurol.,* 1978, 35, 585-9.

Bylsma, FW; Brandt, J; Strauss, ME. Personal and extrapersonal orientation in Huntington's disease patients and those at risk. *Cortex,* 1992, 28, 113-22.

Caltagirone, C; Carlesimo, A; Nocentini, U; Vicari, S. Defective concept formation in parkinsonians is independent from mental deterioration. *J. Neurol. Neurosurg. Psychiatry,* 1989, 52, 334-7.

Canavan, AG; Passingham, RE; Marsden, CD; Quinn, N; Wyke, M; Polkey, CE. Prism adaptation and other tasks involving spatial abilities in patients with Parkinson's disease, patients with frontal lobe lesions and patients with unilateral temporal lobectomies. *Neuropsychologia,* 1990, 28, 969-84.

Couette, M; Bachoud-Levi, AC; Brugieres, P; Sieroff, E; Bartolomeo, P. Orienting of spatial attention in Huntington's Disease. *Neuropsychologia,* 2008, 46, 1391-400.

Crucian, GP; Okun, MS. Visual-spatial ability in Parkinson's disease. *Front. Biosci.,* 2003, 8, s992-7.

Diederich, NJ; Raman, R; Leurgans, S; Goetz, CG. Progressive worsening of spatial and chromatic processing deficits in Parkinson disease. *Arch. Neurol.,* 2002, 59, 1249-52.

Dubois, B; Pillon, B; Sternic, N; Lhermitte, F; Agid, Y. Age-induced cognitive disturbances in Parkinson's disease. *Neurology,* 1990, 40, 38-41.

Fedio, P; Cox, CS; Neophytides, A; Canal-Frederick, G; Chase, TN. Neuropsychological profile of Huntington's disease: patients and those at risk. *Adv. Neurol.,* 1979, 23, 239-55.

Fournet, N; Moreaud, O; Roulin, JL; Naegele, B; Pellat, J. Working memory in medicated patients with Parkinson's disease: the central executive seems to work. *J. Neurol. Neurosurg. Psychiatry,* 1996, 60, 313-7.

Fournet, N; Moreaud, O; Roulin, JL; Naegele, B; Pellat, J. Working memory in medicated patients with Parkinson's disease: the central executive seems to work. *J. Neurol. Neurosurg. Psychiatry,* 1996, 60, 313-7.

Fournet, N; Moreaud, O; Roulin, JL; Naegele, B; Pellat, J. Working memory functioning in medicated Parkinson's disease patients and the effect of withdrawal of dopaminergic medication. *Neuropsychology,* 2000, 14, 247-53.

Freedman, M; Oscar-Berman, M. Spatial and visual learning deficits in Alzheimer's and Parkinson's disease. *Brain Cogn.,* 1989a, 11, 114-26.

Freedman, M; Oscar-Berman, M. Selective delayed response deficits in Parkinson's and Alzheimer's disease. *Arch. Neurol.,* 1989b, 43, 886-90.

Giraudo, MD; Gayraud, D; Habib, M. Visuospatial ability of parkinsonians and elderly adults in location memory tasks. *Brain Cogn.,* 1997, 34, 259-73.

Gómez-Tortosa, E; del Barrio, A; Barroso, T; Garcia Ruiz, PJ. Visual processing disorders in patients with Huntington's disease and asymptomatic carriers. *J. Neurol.,* 1996, 243, 286–92.

Gusella, JF; MacDonald, ME. Huntington's disease: CAG genetics expands neurobiology. *Curr. Opin. Neurobiol.,* 1995, 5, 656-62.

Hietanen, M; Teräväinen, H. The effect of age of disease onset on neuropsychological performance in Parkinson's disease. *J. Neurol. Neurosurg. Psychiatry,* 1988, 51, 244-9.

Hodgson, TL; Dittrich, WH; Henderson, L; Kennard, C. Eye movements and spatial working memory in Parkinson's disease. *Neuropsychologia,* 1999, 37, 927-38.

Hoshi, E. Cortico-basal ganglia networks subserving goal-directed behavior mediated by conditional visuo-goal association. *Front. Neural Circuits,* 2013, 7, 158.

Huber, SJ; Shutteworth, EC; Paulson, GW; Bellchambers, MJG; Clapp, LE. Cortical vs subcortical dementia: neuropsychological differences. *Arch. Neurol.,* 1986, 43, 392-4.

Huntington's disease collaborative group. A novel gene containing a trinucleotide repeat that is expanded and unstable on Huntington's disease chromosomes. *Cell,* 1993, 72, 971-83.

Iwasaki, Y; Kinoshita, M; Ikeda, K; Takamiya, K. Cognitive function in Parkinson's disease: in relation to motor symptoms. *Int. J. Neurosci.,* 1989, 47, 295-300.

Jacobs, DM; Stern, Y; Mayeux, R. Dementia in Parkinson disease, Huntington disease, and other degenerative conditions. In: TE Feinberg, MJ Farah (eds) *Behavioral neurology and neuropsychology*. New York: McGraw-Hill, pp 579-87, 1997.

Jason, GW; Suchowersky, O; Pajurkova, EM; Graham, L; Klimek, ML; Garber, AT; Poirier-Heine, D. Cognitive manifestations of Huntington disease in relation to genetic structure and clinical onset. *Arch. Neurol.,* 1997, 54: 1081-8.

Josiassen, RC; Curry, LM; Mancall, EL. Development of neuropsychological deficits in Huntington's disease. *Arch. Neurol.,* 1983, 40, 791-6.

Kahle, PJ; Neumann, M; Ozmen, L; Muller, V; Jacobsen, H; Schindzielorz, A; Okochi, M; Leimer, U; van Der Putten, H; Probst, A; Kremmer, E; Kretzschmar, HA; Haass, C. Subcellular localization of wild-type and Parkinson's disease-associated mutant α-synuclein in human and transgenic mouse brain. *J. Neurosci.,* 2001, 20, 6365-73.

Kemps, E; Szmalec, A; Vandierendonck, A; Crevits, L. Visuo-spatial processing in Parkinson's disease: evidence for diminished visuo-spatial sketch pad and central executive resources. *Parkinsonism. Relat. Disord.,* 2005, 11, 181-6.

Kilpatrick, GJ; Tilbrook, GS. Memantine. *Curr. Opin. Investig. Drugs,* 2002, 3, 798-806.

Kitada, T; Asakawa, S; Hattori, N; Matsumine, H; Yamamura, Y; Minoshima, S; Yokochi, M; Mizuno, Y; Shimizu, N. Mutations in the parkin gene cause autosomal recessive juvenile parkinsonism. *Nature,* 1998, 392, 605-8.

Krüger, R; Kuhn, W; Müller, T; Woitalla, D; Graeber, M; Kösel, S; Przuntek, H; Epplen, JT; Schöls, L; Riess, O. Ala30Pro mutation in the gene encoding alpha-synuclein in Parkinson's disease. *Nature Genet.*, 1998, 18, 106-8.

Lange, KW; Robbins, TW; Marsden, CD; James, M; Owen, AM; Paul, GM. L-dopa withdrawal in Parkinson's disease selectively impairs cognitive performance in tests sensitive to frontal lobe dysfunction. *Psychopharmacology,* 1992, 107, 394-404.

Laudate, TM; Neargarder, S; Cronin-Golomb, A. Line bisection in Parkinson's disease: investigation of contributions of visual field, retinal vision, and scanning patterns to visuospatial function. *Behav. Neurosci.,* 2013, 127, 151-63.

Lawrence, AD; Watkins, LH; Sahakian, BJ; Hodges, JR; Robbins, TW. Visual object and visuospatial cognition in Huntington's disease: implications for information processing in corticostriatal circuits. *Brain,* 2000, 123, 1349-64.

Le Bras, C; Pillon, B; Damier, P; Dubois, B. At which steps of spatial working memory processing do striatofrontal circuits intervene in humans? *Neuropsychologia,* 1999 37, 83-90.

Lee, AC; Harris, JP; Atkinson, EA; Fowler, MS. Evidence from a line bisection task for visuospatial neglect in left hemiparkinson's disease. *Vision Res.,* 2001, 41, 2677-86.

Lee, FJS; Liu, F. Genetic factors involved in the pathogenesis of Parkinson's disease. *Brain Res. Rev.,* 2008, 58, 354-64.

Leplow, B; Höll, D; Zeng, L; Herzog, A; Behrens, K; Mehdorn, M. Spatial behaviour is driven by proximal cues even in mildly impaired Parkinson's disease. *Neuropsychologia,* 2002, 40, 1443-55.

Levin, BE; Llabre, MM; Visuospatial impairment in Parkinson's disease. *Neurology,* 1991, 41, 365-9.

Levin, BE; Llabre, MM; Weiner, WJ. Cognitive impairments associated with early Parkinson's disease. *Neurology,* 1989, 39, 557-61. Lineweaver, TT; Salmon, DP; Bondi, MW; Corey-Bloom, J. Differential effects of Alzheimer's disease and Huntington's disease on the performance of mental rotation. *J. Int. Neuropsychol.,* 2005, 11, 30-9.

Majerová, V; Kalinčík, T; Laczó, J; Vyhnálek, M; Hort, J; Bojar, M; Růžička, E; Roth, J. Disturbance of real space navigation in moderately advanced but not in early Huntington's disease. *J. Neurol. Sci.,* 2012, 312, 86-91.

Marsden, CD. Slowness of movement in Parkinson's disease. *Mov. Dis.,* 1989, 4 (Suppl 1), S26-S37.

Mendez, MF; Adams, NL; Lewandowsky, KS. Neurobehavioral changes associated with caudate lesions. *Neurology,* 1989, 39, 349-54.

Miller, IN; Neargarder, S; Risi, MM; Cronin-Golomb, A. Frontal and posterior subtypes of neuropsychological deficit in Parkinson's disease. *Behav. Neurosci.,* 2013, 127, 175-83.

Miller, WC; DeLong, MR. Parkinsonian symptomatology: an anatomical and physiological analysis. *Ann. NY Acad. Sci.,* 1988, 515, 287-302.

Mohr, E; Brouwers, P; Claus, JJ; Mann, UM; Fedio, P; Chase, TN. Visuospatial cognition in Huntington's disease. *Mov. Disord.,* 1991, 6, 127-32.

Mohr, E; Juncos, J; Cox, C; Litvan, I; Fedio, P; Chase, TN. Selective deficits in cognition and memory in high-functioning parkinsonian patients. *J. Neurol. Neurosurg. Psychiatry,* 1990a, 53, 603-6.

Mohr, E; Litvan, I; Williams, J; Fedio, Chase, T. Selective deficits in Alzheimer and Parkinsonian dementia: visuospatial function. *Can. J. Neurol. Sci.,* 1990b, 17, 292-7.

Morris, RG; Downes, JJ; Sahakian, BJ; Evenden, JL; Heald, A; Robbins, TW. Planning and spatial working memory in Parkinson's disease. *J. Neurol. Neurosurg. Psychiatry,* 1988, 51, 757-66.

O'Donnell, BF; Blekher, TM; Weaver, M; White, KM; Marshall, J; Beristain, X; Stout, JC; Gray, J; Wojcieszek, JM; Foroud, TM. Visual perception in prediagnostic and early stage Huntington's disease. *J. Int. Neuropsychol. Soc.,* 2008, 14, 446-53.

Owen, AM; Iddon, JL; Hodges, JR; Summers, BA; Robbins, TW. Spatial and non-spatial working memory at different stages of Parkinson's disease. *Neuropsychologia,* 1997, 35, 519-32.

Owen, AM; James, M; Leigh, PN; Summers, BA; Marsden, CD; Quinn, NP; Lange, KW; Robbins, TW. Fronto-striatal cognitive deficits at different stages of Parkinson's disease. *Brain,* 1992, 115, 1727-51.

Owen, AM; Beksinska, M; James, M; Leigh, PN; Summers, BA; Marsden, CD; Quinn, NP; Sahakian, BJ; Robbins, TW. Visuospatial memory deficits at different stages of Parkinson's disease. *Neuropsychologia,* 1993, 31, 627-44.

Partiot, A; Vérin, M; Pillon, B; Teixeira-Ferreira, C; Agid, Y; Dubois, B. Delayed response tasks in basal ganglia lesions in man: further evidence for a striato-frontal cooperation in behavioural adaptation. *Neuropsychologia,* 1996, 34, 709-21.

Pillon, B; Deweer, B; Vidailhet, M; Bonnet, AM; Hahn-Barma, V; Dubois, B. Is impaired memory for spatial location in Parkinson's disease domain specific or dependent on 'strategic' processes? *Neuropsychologia,* 1998, 36, 1-9.

Pillon, B; Ertle, S; Deweer, B; Bonnet, AM; Vidailhet, M; Dubois, B. Memory for spatial location in 'de novo' parkinsonian patients. *Neuropsychologia.* 1997, 35, 221-8.

Pillon, B; Ertle, S; Deweer, B; Sarazin, M; Agid, Y, Dubois, B. Memory for spatial location is affected in Parkinson's disease. *Neuropsychologia,* 1996, 34, 77-85.

Polymeropoulos, MH; Lavedan, C; Leroy, E; Ide, SE; Dehejia, A; Dutra, A; Pike, B; Root, H; Rubenstein, J; Boyer, R; Stenroos, ES; Chandrasekharappa, S; Athanassiadou, A; Papapetropoulos, T; Johnson, WG; Lazzarini, AM; Duvoisin, RC; Di Iorio, G; Golbe, LI; Nussbaum, RL. Mutation in the alpha-synuclein gene identified in families with Parkinson's disease. *Science,* 1997, 276, 2045-7.

Possin, KL; Filoteo, JV; Song, DD; Salmon, DP. Spatial and object working memory deficits in Parkinson's disease are due to impairment in different underlying processes. *Neuropsychology,* 2008, 22, 585-95.

Postert, T; Lack, B; Kuhn, W; Jergas, M; Andrich, J; Braun, B; Przuntek, H; Sprengelmeyer, R; Agelink, M; Büttner, T. Basal ganglia alterations and brain atrophy in Huntington's disease depicted by transcranial real time sonography. *J. Neurol. Neurosurg. Psychiatry,* 1999, 67, 457-62.

Postle, BR; Locascio, JJ; Corkin, S; Growdon, JH. The time course of spatial and object learning in Parkinson's disease. *Neuropsychologia,* 1997, 35, 1413-22.

Raskin, SA; Borod, JC; Wasserstein, J; Bodis-Wollner, I; Coscia, L; Yahr MD. Visuospatial orientation in Parkinson's disease. *Int. J. Neurosci.,* 1990, 51, 9-18.

Rebok, GW; Bylsma, FW; Keyl, PM; Brandt, J; Folstein, SE. Automobile driving in Huntington's disease. *Mov. Dis.,* 1995, 10, 778-87.

Regan, D; Maxner, C. Orientation-selective visual loss in patients with Parkinson's disease. *Brain,* 1987, 110, 415-32.

Robbins, TW; James, M; Owen, AM; Lange, KW; Lees, AJ; Leigh, PN; Marsden, CD; Quinn, NP; Summers, BA. Cognitive deficits in progressive supranuclear palsy, Parkinson's disease, and multiple system atrophy in tests sensitive to frontal lobe dysfunction. *J. Neurol. Neurosurg. Psychiatry,* 1994, 57, 79-88.

Rodríguez-Ferreiro, J; Cuetos, F; Herrera, E; Menéndez, M; Ribacoba, R. Cognitive impairment in Parkinson's disease without dementia. *Mov. Disord.,* 2010, 25, 2136-41.

Roos, RAC. Neuropathology of Huntington's disease. In: PJ Vinken, GW Bruyn, HL Klawans (eds) *Handbook of clinical neurology,* vol 49. Amstersdam: Elsevier, pp 315-26, 1985.

Rosenberg, NK; Sørensen, SA; Christensen, AL. Neuropsychological characteristics of Huntington's disease carriers: a double blind study. *J. Med. Genet.,* 1995, 32, 600–4.

Rouleau, I; Salmon, DP; Butters, N; Kennedy, C; McGuire, K. Quantitative and qualitative analyses of clock drawings in Alzheimer's and Huntington's disease. *Brain Cogn.,* 1992, 18, 70-87.

Sahakian, BJ; Morris, RG; Evenden, JL; Heald, A; Levy, R; Philpot, M; Robbins TW. A comparative study of visuospatial memory and learning in Alzheimer-type dementia and Parkinson's disease. *Brain,* 1988, 111, 695-718.

Sampaio, J; Bobrowicz-Campos, E; André, R; Almeida, I; Faria, P; Januário, C; Freire, A; Castelo-Branco, M. Specific impairment of visual spatial covert attention mechanisms in Parkinson's disease. *Neuropsychologia,* 2011, 49, 34-42.

Shimura, H; Schlossmacher, MG; Hattori, N; Frosch, MP; Trockenbacher, A; Schneider, R; Mizuno, Y; Kosik, KS; Selkoe, DJ. Ubiquitination of a new form of alpha-synuclein by parkin from human brain: implications for Parkinson's disease. *Science,* 2001, 293, 263-9.

Shin, HW; Kang, SY; Sohn, YH. Dopaminergic influence on disturbed spatial discrimination in Parkinson's disease. *Mov. Disord.,* 2005, 20, 1640-3.

Snowden, J; Craufurd, D; Griffiths, H; Thompson, J; Neary, D. Longitudinal evaluation of cognitive disorder in Huntington's disease. *J. Int. Neuropsychol. Soc.*, 2001, 7, 33-44.

Spillantini, MG; Schmidt, ML; Lee, VM; Trojanowski, JQ; Jakes, R; Goedert, M. Alpha-synuclein in Lewy bodies. *Nature,* 1997, 388, 839-40.

Sprengelmeyer, R; Canavan, AGM; Lange, HW; Hömberg, V. Associative learning in degenerative neostriatal disorders: contrasts in explicit and implicit remembering between Parkinson's and Huntington's diseases. *Mov. Dis.,* 1995, 10, 51-65.

Stepánková, K; Ruzicka, E. Object location learning and non-spatial working memory of patients with Parkinson's disease may be preserved in "real life" situations. *Physiol. Res.,* 1998, 47, 377-84.

Stepkina, DA; Zakharov, VV; Yakhno, NN. Cognitive impairments in progression of Parkinson's disease. *Neurosci. Behav. Physiol.,* 2010, 40, 61-7.

Stirk, JA; Foreman, N. Assessment of visual-spatial deficits in patients with Parkinson's disease and closed head injuries using virtual environments. *Cyberpsychol. Behav.,* 2005, 8, 431-40.

Stoffers, D; Berendse, HW; Deijen, JB; Wolters, ECh. Deficits on Corsi's block-tapping task in early stage Parkinson's disease. *Parkinsonism. Relat. Disord.,* 2003, 10, 107-11.

Taylor, AE; Saint-Cyr, JA; Lang, AE. Frontal lobe dysfunction in Parkinson's disease: the cortical focus of neostriatal outflow. *Brain,* 1986, 109, 845-83.

Taylor, AE; Saint-Cyr, JA; Lang, AE. Parkinson's disease: cognitive changes in relation to treatment response. *Brain,* 1987, 110, 35-51.

Taylor, HG; Hansotia, P. Neuropsychological testing of Huntington's patients. Clues to progression. *J. Nerv. Ment. Dis.,* 1983, 171, 492-6.

Tröster, AI; Butters, N; Salmon, DP; Cullum, CM; Jacobs, D; Brandt, J; White, RF. The diagnostic utility of savings scores: differentiating Alzheimer's and Huntington's diseases with the logical memory and visual reproduction tests. *J. Clin. Exp. Neuropsychol.,* 1993, 15, 773-88.

Tsai, C-H; Lu, C-S; Hua, M-S, Lo, W-L; Lo, S-K. Cognitive dysfunction in early onset parkinsonism. *Acta Neurol. Scand.,* 1994, 89, 9-14.

Yágüez, L; Lange, HW; Hömberg, V. Differential effect of Huntington's and Parkinson's diseases in programming motor sequences of varied lengths. *J. Neurol.,* 2006, 253, 186-93.

Wallesch, CW; Karnath, HO; Papagno, C; Zimmermann, P; Deuschl, G; Lücking, CH. Parkinson's disease patient's behaviour in a covered maze learning task *Neuropsychologia,* 1990, 28, 839-49.

Wang, M; Hattori, N; Matsumine, H; Kobayashi, T; Yoshino, H; Morioka, A; Kitada, T; Asakawa, S; Minoshima, S; Shimizu, N; Mizuno, Y. Polymorphism in the parkin gene in sporadic Parkinson's disease. *Ann. Neurol.,* 1999, 45, 655-8.

Weiner, WJ. Parkinson's disease. In: WJ Weiner, CG Goetz (eds) *Neurology for the non-neurologist*. Philadelphia: JP Lippincott, 2nd ed, pp 109-20, 1989.

In: Spatial Orientation in Humans and Animals
Editor: Robert Lalonde
ISBN: 978-1-63463-001-6
© 2014 Nova Science Publishers, Inc.

Chapter 7

SPATIAL DEFICITS IN ANIMAL MODELS OF BASAL GANGLIA DISORDERS

Robert Lalonde*
Université de Rouen, Dépt. Psychologie,
Mont-Saint-Aignan Cedex France

1. MORRIS WATER MAZE

1.1. Dopamine-Restricted Lesions and Receptor Blockers

The neurochemical basis of striatal contributions to maze behavior (Table 1) has been examined with selective catecholamine neurotoxins such as 6-hydroxy-dopamine (6-OHDA). When combined with peripheral pretreatment of noradrenaline reuptake blockers such as desipramine and talsupram, intracebral 6-OHDA administration selectively depletes dopamine without affecting noradrenaline. Because of gastrointestinal distress sometimes reported with desipramine, the monoamine oxidase inhibitor, pargyline, has sometimes been preferred (Whishaw and Dunnett, 1985).

The acquisition of the hidden platform subtask of the Morris water maze (chapter 2) was slowed down in rats after 6-OHDA+ desipramine-induced lesions in the substantia nigra pars compacta (SNc) (Andringa et al., 2000), source of ascending dopamine fibers to the dorsal striatum (chapter 3). In addition, 6-OHDA injections in the SNc impaired the visible platform subtask (Ferro et al., 2005), which implies a loss in motivation or visuomotor control. These effects are not limited to adult rats, since intraventricular (Archer et al., 1988, 2002) or intracisternal (Rex and Fink, 2004) injections of 6-OHDA along with a noradrenaline uptake blocker during the neonatal period delayed acquisition of the hidden platform task during the juvenile-adult stage. On the contrary, intra-nasal (Prediger et al., 2006) or intra-SNc (Da Cunha et al., 2002, 2006, 2007) administration of a second dopamine-depleting neurotoxin, 1-methyl-4-phenyl-l,2,3,6-tetrahydropyridine (MPTP), had no effect during acquisition of the submerged platform task.

---

* Email: robert.lalonde@univ-rouen.fr

Table 1. Spatial deficits after dopamine-restricted lesions on the Morris water maze and the radial-arm maze

Neurotoxins	Brain regions	Tests	References
6-OHDA	substantia nigra pars compacta	Morris water maze	Andringa et al., 2000; Ferro et al., 2005
MPTP	substantia nigra pars compacta	Morris water maze	Da Cunha et al., 2002, 2007; Kumar et al., 2009; Luchtman et al., 2012; Miyoshi et al., 2002; Perry et al., 2004; Reksidler et al., 2007
6-OHDA	dorsal striatum	Morris water maze	Braun et al., 2012; Hagan et al., 1983; Tadaiesky et al., 2008
botulinum neurotoxin-A	dorsolateral striatum	Morris water maze	Holzmann et al., 2012
6-OHDA	hippocampus	Morris water maze	Gasbarri et al., 1996
6-OHDA	lateral septum	radial-arm maze	Simon et al., 1986

Yet MPTP injections in the SNc slowed down the acquisition of a simpler version with constant rather than variable starting points (Da Cunha et al., 2006) and impaired swimming towards the visible platform (Da Cunha et al., 2002, 2007; Kumar et al., 2009; Luchtman et al., 2012; Miyoshi et al., 2002; Perry et al., 2004; Reksidler et al., 2007), the latter result found as well when the substance is administered through the intra-nasal route (Castro et al., 2013; Prediger et al., 2006). Negative results on the hidden platform task probably reflect an increase in motivation relative to the visible platform task due to a higher degree of perceived danger.

Dopamine neurotoxins affect the Morris task when administered throughout the nigrostriatal pathway. Indeed, the acquisition and performance of hidden and visible platform tasks were deficient after 6-OHDA+ pargyline injections in the ascending nigrostriatal pathway (Whishaw et al., 1985). The deleterious impact of this neurotoxin on the visible platform task was subsequently confirmed when injected into the nigrostriatal pathway (Mura and Feldon, 2003). In addition, 6-OHDA slowed down acquisition of the hidden (Braun et al., 2012) or visible (Tadaiesky et al., 2008) platform tasks when injected in the dorsal striatum, as did 6-OHDA+ desipramine in the same brain region (Hagan et al., 1983). Likewise, the hidden platform task was delayed after botulinum neurotoxin-A injections in the dorsolateral striatum (Holzmann et al., 2012).

In addition to the nigrostriatal pathway, dopamine appears to influence the behavioral task at the level of the mesolimbic pathway. Indeed, submerged platform and probe tests, but not the visible platform test, were impaired after 6-OHDA+ desipramine injections in the CA1 subfield of the hippocampal formation or in the subiculum (Gasbarri et al., 1996). On the contrary, 6-OHDA+ desipramine injections in either nucleus accumbens or prefrontal cortex had no effect on hidden or visible platform tasks (Hagan et al., 1983).

In addition to the standard protocol, rats have been examined in a working memory version with different platform locations (chapter 2). 6-OHDA injections in the SNc impaired this version (Ferro et al., 2005), as did MPTP administered through intranasal (Prediger et al., 2006) or intra-SNc (Ferro et al., 2005; Miyoshi et al., 2002) routes.

1.2. Mutant Models of Parkinson's Disease

Several murine mutants have been engineered to mimic Parkinson's disease (Fleming et al., 2005), including A30P or A53T mutations and wild-type overexpression of human *SNCA* (Table 2). A *Pitx3*/ak murine knockout of the *Pitx3* gene, encoding paired-like homeodomain transcription factor 3, results in brain dopamine depletion, but its use is limited in regard to spatial tasks because of a reduction in visual acuity (Ardayfio et al., 2008). Parkinsonian neuropathology is also a feature of mice deleted for *Tfam* (mitochondrial transcription factor A) (Li et al., 2013).

Table 2. Spatial deficits in murine models of Parkinson's disease

Gene	Mutation	Tests	References
SNCA transgenic	A30P	Morris water maze	Freichel et al., 2007
SNCA transgenic	A53T	spontaneous alternation	Paumier et al., 2013
SNCA transgenic	Y39C	Morris water maze	Zhou et al., 2008
SNCA transgenic	wild-type	Morris water maze, spatial reversal, spontaneous alternation	Magen et al., 2012; Nuber et al., 2008
Park2 knockout	not applicable	Morris water maze, spontaneous alternation	Itier et al., 2003; Zhu et al., 2007
Pitx3 knockout	not applicable	left-right discrimination	Ardayfio et al., 2008
Tfam knockout	not applicable	Lashley III maze	Li et al., 2013

Relative to wild-type controls, the acquisition of the hidden but not the visible platform task was delayed in *SNCA*/Y39C transgenic mice (Zhou et al., 2008). In contrast, the probe trial, not acquisition of the hidden platform task, was impaired in *SNCA*/A30P (Freichel et al., 2007) and *SNCA*/wild-type (Nuber et al., 2008) mutants. Moreover, although the acquisition of the hidden platform subtest was slowed down in a *Park2* knockout with deletion of exon 2 (Zhu et al., 2007), this was not so in another mutant with the same deletion (Perez and Palmiter, 2005). In general, the impact of genetic modifications on the Morris water maze tasks appears more minor than that reported after dopamine neurotoxins, probably because of insufficient dopamine depletions. Likewise, transgenic mutants with murine *SNCA* mutations show less neuropathology than those with human *SNCA* mutations (Rieker al, 2012).

1.3. Electrolytic or Excitotoxic Lesions

In addition to dopamine, the more general role of the striatum on the Morris water maze task has been examined. Electrolytic lesions of the dorsal striatum slowed down the acquisition of hidden and visible platform tasks (Devan et al., 1996). The acquisition of the submerged platform task was also slowed down after excitotoxic lesions of the dorsal striatum (Ayalon et al., 2004; Block et al., 1993; Emerich et al., 1997; Garside et al., 1996;

Joel et al., 1998; Sagratella et al., 1994; Whishaw et al., 1987). In addition to acquisition, excitotoxic lesions of the striatum impaired the performance of the already acquired task (Whishaw et al., 1987). Likewise, the acquisition of the hidden platform task was slowed down after electrolytic lesions of the lateral pallidum (Jeljeli et al., 1999), a brain region receiving afferent fibers from the striatum (chapter 3). When subregions were examined, electrolytic lesions of dorsomedial striatum slowed down the acquisition of submerged and visible platform tests (Devan et al., 1999), as did excitotoxic lesions of medial striatum on acquisition of hidden (Furtado and Mazurek, 1996) and dorsolateral striatum on visible (Martel et al., 2007) platform tasks, respectively.

As shown above regarding dopaminergic lesions, the mesolimbic system also appears involved. Indeed, the acquisition of the hidden platform task was slowed down after electrolytic (Sutherland and Rodriguez, 1989) or excitotoxic (Annett et al., 1996) lesions of the nucleus accumbens. Moreover, the probe trial was impaired after excitotoxic lesions of this region (Annett et al., 1996). However, the acquisition of the hidden platform task was not affected after electrolytic (Thifault et al., 1998) or lidocaine-induced (Floresco et al., 1996) lesions restricted to only its medial (shell) part.

1.4. Mutant Models of Huntington's Disease

Huntington's disease is caused by an expansion of more than 39 cytosine-adenine-guanine (CAG) repeats in the *HD* gene, encoding huntingtin (chapter 6). Several mutant models have been engineered to mimic this disease (Heng et al., 2008), including R6/1 mice carrying 116 CAG repeats of exon 1 of the human *HD* gene and R6/2 mice carrying 144 CAG repeats, which, as in human cases, exhibit involuntary movements (Mangiarini et al., 1996). Also as in human cases, the numbers increase with succeeding generations, so that a few years later R6/1 mice averaged 124 CAG repeats (Brooks et al., 2012b). R6/2 mutants had reduced regional brain metabolism as measured by cytochrome oxidase activity (Tabrizi et al., 2000) and showed evidence of neuronal cell death in neocortex and striatum (Davies et al., 1999). Other models of Huntington's disease comprise YAC128 mutant mice carrying 128 CAG repeats of the full-length *HD* gene on a yeast artificial chromosome (Slow et al., 2003), *Hd*/Q111 knockin mice with 111 CAG repeats (Giralt et al., 2012), and a rat model carrying a truncated huntingtin cDNA fragment with 51 CAG repeats under control of the native rat *Hd* promoter (von Hörsten et al., 2003). YAC128 mice (Slow et al., 2003; van Raamsdonk et al., 2007) and *HD* transgenic rats (Cao et al., 2006) exhibit striatal atrophy and progressive motor dysfunctions.

R6/1 mice were slower than non-transgenic controls in acquiring the hidden platform task and deficient during the probe trial (Brooks et al., 2012b). Impairments in hidden platform (Lione et al., 1999; Lüesse et al., 2001; Murphy et al., 2000) and probe (Lione et al., 1999; Murphy et al., 2000; Wood et al., 2010) tasks are also a feature of R6/2 mice. Despite some awkwardness in swimming (Carter et al., 1999), visuomotor guidance towards the visible platform was not affected in R6/2 mice (Lione et al., 1999; Lüesse et al., 2001). Likewise, YAC128 mutants travelled further and took longer than controls before finding the hidden platform (Brooks et al., 2012a). Although the behavior of *Hd*/Q111 knockins was the same as controls in acquiring hidden and visible platform tasks, they were impaired in the probe test of long-term memory (Giralt et al., 2012).

In an adaptation of the Morris water maze, R6/2 mice were trained to swim from a fixed starting point to a cued platform (Ciamei and Morton, 2009). During the probe trial, the pool was rotated by 180 degrees, and mice chose between a hidden platform located where the cued platform had been (place learning) or one located in the opposite quadrant (cued learning). At 2 months of age, R6/2 mice were mainly driven by a striatal-mediated cue-based strategy but one month later by a hippocampal place-based strategy. Likewise, *HD* transgenic rats were impaired in both paradigms (Kirch et al., 2013).

2. Multi-Unit Mazes, Hole-Board, Object Location

Fewer experiments have been conducted with other reference memory tasks such as the multi-unit maze, the Barnes hole-board, and object location (chapter 2).

2.1. Dopamine-restricted lesions

Rats injected with 6-OHDA in the dorsal striatum were slower in acquiring the Cincinnati water maze (Braun et al., 2012). Mutant mice with a deletion of *Tfam* were slower in acquiring the Barnes hole-board task (Li et al., 2013). In contrast, *Park2* null mutant mice were not affected relative to controls in a water version of the Lashley III maze (Perez and Palmiter, 2005).

2.2. Electrolytic or Excitotoxic Lesions

Excitotoxic lesions in dorsomedial or dorsolateral striatum delayed acquisition of a 14-unit T-maze motivated by shock escape (Pistell et al., 2009). No such effect was found after electrolytic lesions of the dorsomedial striatum in the Hebb-Williams maze (Chorover and Gross, 1963).

2.3. Mutant Models of Huntington's Disease

The R6/2 mutant mouse model of Huntington's disease was slower than non-transgenic controls in finding the goal area of the water-filled Lashley III maze (Wood et al., 2011). The very similar R6/1 model was deficient in acquiring the Barnes hole-board task (Nithianantharajah et al., 2008) as were *HD* rats for the object location task (Zeef et al., 2012). It therefore appears that several reference memory tasks are vulnerable to the effects of Huntington's disease transgenesis.

3. Left-Right Discrimination Learning

3.1. Dopamine-Restricted Lesions

Left-right discrimination and reversal learning (chapter 2) have been examined in the context of dopaminergic pathways in rats. The acquisition of left-right discrimination learning was not slowed down after 6-OHDA infusions in the dorsomedial striatum for food reward (Lex et al., 2011) or water escape (Haik et al., 2008). However, spatial reversal training was impaired by this lesion (Haik et al., 2008), as well as dopamine depletion in the nucleus accumbens (Taghzouti et al., 1985). The latter also affected original training. The acquisition of both tasks was also delayed after 6-OHDA+ desipramine-induced lesions in the lateral septum (Simon et al., 1986). Paradoxical results were obtained when 6-OHDA was injected in the lateral septum of mice, in that the treatment facilitated acquisition and reversal of spatial discrimination learning (Galey et al., 1985). It remains to be determined whether these opposite results are due to a species difference or to a difference in dopamine levels.

3.2. Mutant Models of Parkison's Disease

Evidence for a role of dopamine in such tasks has been extended to genetically modified mice. *SNCA*/wild-type mutants were slower than non-mutants to acquire spatial reversal learning but not the original right-left discrimination (Magen et al., 2012). Relative to wild-type controls and mutant mice with retinal degeneration, *Pitx3*/ak mice acquired more slowly the left-right discrimination task in a water-filled T-maze (Ardayfio et al., 2008). However, no such result occurred in a dry T-maze, implicating a reduction in motivation rather than cognitive abilities as the underlying cause of the water maze deficit.

3.3. Electrolytic or Excitotoxic Lesions

Several experiments indicate a role for the basal ganglia on food-motivated left-right learning in rats. Indeed, the acquisition of left-right discrimination and spatial reversals was slowed down after electrolytic lesions of the dorsal striatum (Mitchell and Hall, 1988; Mitchell et al., 1985) or only its dorsomedial part (Cook and Kesner, 1988). Deficits in both tasks were also reported after excitotoxic lesions of the dorsomedial striatum (Dunnett and Iversen, 1981). Some researchers have found slowed reversal but not original training after excitotoxic (Pisa and Cyr, 1990) or electrolytic (Kolb, 1977) lesions of the medial striatum. However, Schwartzbaum and Donovick (1968) found no effect on either task. The variability in results may be due to the extent of damage or to spontaneous post-operative recovery. Nevertheless, the acquisition of spatial reversals was retarded after aspiration lesions of the dorsomedial striatum in cats (Olmstead et al., 1976). As shown for dopamine-restricted lesions, the mesolimbic system appears involved in that acquisition of left-right discrimination learning and spatial reversal was delayed after excitotoxic lesions of the nucleus accumbens (Annett et al., 1989).

In addition to acquisition, the basal ganglia exert an influence on the performance of spatial reversals, as documented after electrolytic lesions of anterior (Divac, 1972) or dorsomedial (Divac, 1971) striatum. Electrolytic lesions of the dorsomedial striatum also impaired the performance of finding one baited arm among eight in the radial-arm maze (Cook and Kesner, 1988).

Left-right discrimination learning has also been examined with the shock avoidance paradigm. The acquisition of spatial reversals motivated by fear of footshocks was slowed down after electrolytic lesions of either the dorsomedial striatum or substantia nigra (Thompson and Yang, 1982). Likewise, excitotoxic lesions of the dorsal striatum delayed left-right discrimination and reversal learning when water escape was used as motivation (Ayalon et al., 2004).

3.4. Mutant Models of Huntington's Disease

The YAC128 mouse model was assessed in a water T-maze set-shifting procedure with spatial learning partly involved (Brooks et al., 2012c). The task included six stages: directional learning, directional learning reversal, light discrimination, light discrimination reversal, return to place learning, and a maze rotation spatial learning test. Rule changes from place learning to light discrimination and back constitute extra-dimensional shifts. The results showed deficits in directional learning to light discrimination extra-dimensional set-shifting.

4. Radial-Arm Maze

In contrast to the tasks reviewed above, the radial-arm maze task reflects trial-dependent working memory in that the solution differs from one trial to the next, though reference memory may be examined concomitantly by not baiting some of the arms (chapter 2). The effects of experimental lesions on the radial-arm maze task are depicted in Tables 1, 3, and 4.

Table 3. Spatial deficits after basal ganglia-restricted lesions on the Morris water maze and the radial-arm maze

Lesion method	Brain regions	Tests	References
electrolytic	dorsal striatum	Morris water maze	Devan et al., 1996
excitotoxic	dorsal striatum	Morris water maze	Ayalon et al., 2004; Block et al., 1993; Emerich et al., 1997; Garside et al., 1996; Joel et al., 1998; Sagratella et al., 1994; Whishaw et al., 1987
electrolytic	dorsomedial striatum	Morris water maze	Devan et al., 1999
excitotoxic	medial striatum	Morris water maze	Furtado and Mazurek, 1996
excitotoxic	dorsolateral striatum	Morris water maze	Martel et al., 2007
electrolytic	lateral pallidum	Morris water maze	Jeljeli et al., 1999
electrolytic	nucleus accumbens	Morris water maze	Sutherland and Rodriguez, 1989
excitotoxic	nucleus accumbens	Morris water maze	Annett et al., 1996

Table 3. (Continued)

electrolytic	dorsomedial striatum	radial-arm maze	Colombo et al., 1989
excitotoxic	striatum	radial-arm maze	Haik et al., 2000; Lescaudron et al., 2003; Shear et al., 1998a,b
electrolytic	lateral pallidum	radial-arm maze	Ennaceur, 1998
electrolytic	nucleus accumbens	radial-arm maze	Gal et al., 1997
excitotoxic	nucleus accumbens	radial-arm maze	Schacter et al., 1989

Table 4. Spatial deficits in murine models of Huntington's disease

Mutation type	CAG repeats	Tests	References
R6/1 transgenic mouse	116	Morris water maze, Barnes hole-board, spontaneous alternation	Brooks et al., 2012b: Dowie et al., 2010: Nithianantharajah et al., 2008
R6/2 transgenic mouse	144	Morris water maze, Lashley III maze, spontaneous alternation	Lione et al., 1999; Lüesse et al., 2001; Murphy et al., 2000; Wood et al., 2010, 2011
YAC128 transgenic mouse	128	Morris water maze	Brooks et al., 2012a
Hd knockin mouse	111	Morris water maze	Giralt et al., 2012
HD transgenic rat	51	object location radial-arm maze	Nguyen et al., 2006; von Hörsten et al., 2003; Zeef et al., 2012

4.1. Dopamine-Restricted Lesions

Rats injected with botulinum neurotoxin-A in the dorsolateral striatum were not affected during acquisition of the radial-arm maze task (Holzmann et al., 2012). In contrast, the acquisition of the radial-arm maze task was impaired after 6-OHDA+ desmethylimipramine-induced lesions of the lateral septum (Simon et al., 1986).

4.2. Electrolytic or Excitotoxic Lesions

Post-operative performance of reference memory in the radial-arm maze task motivated by food reward was impaired after electrolytic lesions of the dorsomedial striatum (Colombo et al., 1989). In contrast, such lesions did not impair performance of the working memory component (Becker et al., 1980; Colombo et al., 1989; Cook and Kesner, 1988). In contrast, excitotoxic lesions of the striatum impaired working memory in a water maze paradigm (Haik et al., 2000; Lescaudron et al., 2003; Shear et al., 1998a,b), possibly because the motivation to escape from drowning, more urgent than the motivation to obtain food, incites rats to adopt a win-stay rather than a win-shift strategy (chapter 2). Nevertheless, such lesions impaired performance in a version requiring the finding of food pellets in adjacent arms (Colombo et al., 1989; Cook and Kesner, 1988). Likewise, electrolytic lesions of the dorsomedial striatum slowed down acquisition of an egocentric version of the radial-arm maze task (Potegal, 1969).

Moreover, electrolytic lesions of the same region delayed acquisition of a win-stay version signaled by a light source (Packard et al., 1989).

Basal ganglia influence extends to striatal efferent structures, since electrolytic lesions of the lateral pallidum slowed down acquisition of the radial-arm maze task (Ennaceur, 1998). The effect of lesioning the mesolimbic system has also been examined. The acquisition of the radial-arm maze task was retarded after electrolytic lesions of the shell part of the nucleus accumbens for both working and reference memory (Gal et al., 1997). But in another study, only the acquisition of reference memory was retarded after excitotoxic injections of the nucleus accumbens (Schacter et al., 1989).

4.3. Mutant Models of Huntington's Disease

HD transgenic rats were impaired in the radial-arm maze task (von Hörsten et al., 2003), with impairments seen both in regard to working and reference memory (Nguyen et al., 2006).

5. Spatial Delayed Response

Like the radial-arm maze task, spatial delayed response reflects trial-dependent working memory (chapter 2). To our knowledge, the evaluation of spatial delayed response following dopamine depletions has only been examined in monkeys. The performance of spatial delayed response was impaired after systemic injections of MPTP (Fernández-Ruiz et al., 1999). This result may be due to depleted dopamine levels in the prefrontal cortex, since the acquisition of spatial delayed response was slowed down after 6-OHDA+talsupram infusions in prefrontal cortex (Roberts et al., 1994).

6. Left-Right Alternation or Non-Matching-to-Place Learning

Like radial-arm maze and delayed response tasks, left-right alternation and non-matching-to-place tasks reflects trial-dependent working memory (chapter 2).

6.1. Dopamine-Restricted Lesions

The acquisition of reinforced alternation was slowed down after 6-OHDA injected unilaterally in the nigro-striatal pathway of rats (Ramos-Moreno et al., 2012). The performance of reinforced alternation was impaired in mice injected peripherally with MPTP (Tanila et al., 1998) and in rats injected with the same toxin in the substantia nigra (Braga et al., 2005). Intra-nigral infusions of MPTP also delayed acquisition of the non-matching-to-place task (Ho et al., 2011; Hsieh et al., 2002).

6.2. Electrolytic and Excitotoxic Lesions

The acquisition of left-right alternation learning was slowed down after electrolytic (Mogensen et al., 1987) or excitotoxic (Furtado and Mazurek, 1996) lesions of the medial striatum. The same effect was reported after electrolytic lesions of the ventral tegmental area (Stam et al., 1989). Likewise, the performance of left-right alternation learning was impaired after electrolytic (Divac, 1972) or excitotoxic (Divac et al., 1978, Wikmark et al., 1973) lesions of the medial striatum. However, Colombo et al. (1989) found no such effect after electrolytic lesions of the dorsomedial striatum.

In addition to turning sides inside a maze, the role of the basal ganglia has been examined with lever-pressing responses. The acquisition and performance of alternating bar-presses were impaired after electrolytic lesions of the dorsomedial striatum (Chorover and Gross, 1963; Gross et al., 1965).

Likewise, the performance of alternating bar-presses declined after excitotoxic lesions in striatum (Dunnett et al., 1999). However, no such effect was discovered after electrolytic lesions of the medial striatum during the performance of alternating nose-pokes (Mogensen et al., 1987). Instead, the performance of the task was impaired after excitotoxic lesions of the ventral tegmental area (Kessler and Markowitsch, 1981).

6.4. Mutant Models of Huntington's Disease

For delayed alternation with lever-pressing responses, there was no effect among *HD* transgenic rats against wild-type littermates (Fielding et al., 2012).

7. Spontaneous Alternation

7.1. Dopamine-Restricted Lesions

Rats injected unilaterally with 6-OHDA in the SNc had lower spontaneous alternation rates than controls (Ciobica et al., 2012). The same effect was found after unilateral (Kopalli et al., 2013) or bilateral (Hefco et al., 2003) infusions of 6-OHDA+ desipramine in SNc or bilateral infusions in ventral tegmental area (Hefco et al., 2003). However, Pioli et al. (2008) found lower spontaneous alternation rates after bilateral 6-OHDA+desipramine injections in ventral tegmental area but not SNc.

Dopaminergic depletion in the nucleus accumbens also reduced spontaneous alternation rates in rats (Taghzouti et al., 1985). In contrast, mice injected with 6-OHDA in the lateral septum had higher spontaneous alternation rates (Galey et al., 1985).

7.2. Mutant Models of Parkinson's Disease

Compared with age-matched wild-type littermates, *SNCA*/wild-type mutants made fewer spontaneous alternations than controls (Magen et al., 2012). Spontaneous alternation rates also declined in mice with the exon 3 deletion of *Park2* (Itier et al., 2003), but not in those with the exon 2 deletion (Perez and Palmiter, 2005). In a version measuring time spent in a novel arm of a Y-maze instead of rates, *SNCA*/53T mutants were impaired (Paumier et al., 2013).

7.3. Electrolytic and Excitotoxic Lesions

Electrolytic lesions of the dorsal striatum reduced their spontaneous alternation rates in rats (Divac et al., 1975), the same deficit reported after excitotoxic lesions of medial striatum (Furtado and Mazurek, 1996) as well as substantia nigra or ventral tegmental area (Hefco et al., 2003).

7.4. Mutant Models of Huntington's Disease

As expected from studies in rats with dorsal striatal lesions, R6/1 (Dowie et al., 2010) and R6/2 (Lione et al., 1999) mice were impaired in spontaneous alternation testing relative to non-transgenic controls.

REFERENCES

Ardayfio, P; Moon, J; Leung, KK; Youn-Hwang, D; Kim, KS. Impaired learning and memory in Pitx3 deficient aphakia mice, a genetic model for striatum-dependent cognitive symptoms in Parkinson's disease. *Neurobiol. Dis.,* 2008, 31, 406-12.

Andringa, G; van Oosten, RV; Unger, W; Hafmans, TG; Veening, J; Stoof, JC; Cools, AR. Systemic administration of the propargylamine CGP 3466B prevents behavioural and morphological deficits in rats with 6-hydroxydopamine-induced lesions in the substantia nigra. *Eur. J. Neurosci.,* 2000, 12, 3033-43.

Annett, LE; McGregor, A; Robbins, TW. The effects of ibotenic acid lesions of the nucleus accumbens on spatial learning and extinction in the rat. *Behav. Brain Res.,* 1989, 31, 231-42.

Archer, T; Danysz, W; Fredriksson, A; Jonsson, G; Luthman, J; Sundström, E; Teiling, A. Neonatal 6-hydroxydopamine-induced dopamine depletions: motor activity and performance in maze learning. *Pharmacol. Biochem. Behav.,* 1988, 31, 357-64.

Archer, T; Palomo, T; Fredriksson, A. Functional deficits following neonatal dopamine depletion and isolation housing: circular water maze acquisition under pre-exposure conditions and motor activity. *Neurotox. Res.,* 2002, 4, 503-22.

Ayalon, L; Doron, R; Weiner, I; Joel D. Amelioration of behavioral deficits in a rat model of Huntington's disease by an excitotoxic lesion to the globus pallidus. *Exp. Neurol.,* 2004, 186, 46-58.

Becker, JT; Walker, JA; Olton, DS. Neuroanatomical bases of spatial memory. *Brain Res.,* 1980, 200, 307-20.

Block, F; Kunkel, M; Schwarz, M. Quinolinic acid lesion of the striatum induces impairment in spatial learning and motor performance in rats. *Neurosci. Lett.*, 1993, 149, 126-8.

Braga, R; Kouzmine, I; Canteras, NS; Da Cunha, C. Lesion of the substantia nigra, pars compacta impairs delayed alternation in a Y-maze in rats. *Exp. Neurol.*, 2005, 192, 134-41.

Braun, AA; Graham, DL; Schaefer, TL; Vorhees, CV; Williams, MT. Dorsal striatal dopamine depletion impairs both allocentric and egocentric navigation in rats. *Neurobiol. Learn. Mem.*, 2012, 97, 402-8.

Brooks, S; Higgs, G; Janghra, N; Jones, L; Dunnett, SB. Longitudinal analysis of the behavioural phenotype in YAC128 (C57BL/6J) Huntington's disease transgenic mice. *Brain Res. Bull.*, 2012a, 88, 113-20.

Brooks, SP; Janghra, N; Workman, VL; Bayram-Weston, Z; Jones, L; Dunnett, SB. Longitudinal analysis of the behavioural phenotype in R6/1 (C57BL/6J) Huntington's disease transgenic mice. *Brain Res. Bull.*, 2012b, 88, 94-103.

Brooks, SP; Janghra, N; Higgs, GV; Bayram-Weston, Z; Heuer, A; Jones, L; Dunnett, SB. Selective cognitive impairment in the YAC128 Huntington's disease mouse. *Brain Res. Bul.l*, 2012c, 88, 121-9.

Cao, C; Temel, Y; Blokland, A; Ozen, H; Steinbusch, HW; Vlamings, R; Nguyen, HP; von Hörsten, S; Schmitz, C; Visser-Vandewalle, V. Progressive deterioration of reaction time performance and choreiform symptoms in a new Huntington's disease transgenic rat model. *Behav. Brain Res.*, 2006, 170, 257-61.

Carter, RJ; Lione, LA; Humby, T; Mangiarini, L; Mahal, A; Bates, GP; Dunnett, SB; Morton, AJ. Characterization of progressive motor deficits in mice transgenic for the human Huntington's disease mutation. *J. Neurosci.*, 1999, 19, 3248-57.

Castro, AA; Wiemes, BP; Matheus, FC; Lapa, FR; Viola, GG; Santos, AR; Tasca, CI; Prediger, RD. Atorvastatin improves cognitive, emotional and motor impairments induced by intranasal 1-methyl-4-phenyl-1,2,3,6-tetrahydropyridine (MPTP) administration in rats, an experimental model of Parkinson's disease. *Brain Res.*, 2013, 1513, 103-16.

Chorover, SL; Gross, CG. Caudate nucleus lesions: behavioral effects in the rat. *Science*, 1963, 141, 826-7.

Ciamei, A; Morton, AJ. Progressive imbalance in the interaction between spatial and procedural memory systems in the R6/2 mouse model of Huntington's disease. *Neurobiol. Learn. Mem.*, 2009, 92, 417-28.

Ciobica, A; Padurariu, M; Hritcu, L. The effects of short-term nicotine administration on behavioral and oxidative stress deficiencies induced in a rat model of Parkinson's disease. *Psychiatr. Danub.*, 2012, 24, 194-205.

Colombo, PJ; Davis, HP; Volpe, BT. Allocentric spatial and tactile memory impairments in rats with dorsal caudate lesions are affected by preoperative behavioral training. *Behav. Neurosci.*, 1989, 103, 1242-50.

Cook, D; Kesner RP. Caudate nucleus and memory for egocentric localization. *Behav. Neural Biol.*, 1988, 49, 332-43.

Da Cunha, C; Angelucci, ME; Canteras, NS; Wonnacott, S; Takahashi, RN. The lesion of the rat substantia nigra pars compacta dopaminergic neurons as a model for Parkinson's disease memory disabilities. *Cell. Mol. Neurobiol.*, 2002, 22, 227-37.

Da Cunha, C; Silva, MH; Wietzikoski, S; Wietzikoski, EC; Ferro, MM; Kouzmine, I; Canteras, NS. Place learning strategy of substantia nigra pars compacta-lesioned rats. *Behav. Neurosci.,* 2006, 120, 1279-84.

Da Cunha, C; Wietzikoski, S; Wietzikoski, EC; Silva, MH; Chandler, J Jr; Ferro, MM; Andreatini, R; Canteras, NS. Pre-training to find a hidden platform in the Morris water maze can compensate for a deficit to find a cued platform in a rat model of Parkinson's disease. *Neurobiol. Learn. Mem.,* 2007, 87, 451-63.

Davies, SW; Turmaine, M; Cozens, BA; Raza, AS; Mahal, A; Mangiarini, L; Bates, GP. From neuronal inclusions to neurodegeneration: neuropathological investigation of a transgenic mouse model of Huntington's disease. *Trans. Roy. Soc. London B. Biol. Sci.,* 1999, 354, 971-79.

Decamp, E; Tinker, JP; Schneider, JS. Attentional cueing reverses deficits in spatial working memory task performance in chronic low dose MPTP-treated monkeys. *Behav. Brain Res.,* 2004, 152, 259-62.

Devan, BD; Goad, EH; Petri, HL. Dissociation of hippocampal and striatal contributions to spatial navigation in the water maze. *Neurobiol. Learn. Mem.*, 1996, 66, 305-23.

Devan, BD; McDonald, RJ; White, NM. Effects of medial and lateral caudate-putamen lesions on place- and cue-guided behaviors in the water maze: relation to thigmotaxis. *Behav. Brain Res.,* 1999, 100, 5-14.

Divac, I. Frontal lobe system and spatial reversal in the rat. *Neuropsychologia,* 1971, 9, 175-83.

Divac, I. Neostriatum and functions of prefrontal cortex. *Acta Neurobiol. Exp.,* 1972, 32, 461-77.

Divac, I; Wiskmar, RGE; Gade, A. Spontaneous alternation in rats with lesions of the frontal lobes: an extension of the frontal lobe syndrome. *Physiol. Psychol.,* 1975, 3, 39-42.

Divac, I; Markowitsch, HJ; Pritzel, M. Behavioral and anatomical consequences of small intrastriatal injections of kainic acid in the rat. *Brain Res.,* 1978, 151, 523-32.

Dunnett, SB; Iversen, SD. Learning impairments following selective kainic acid-induced lesions within the neostriatum of rats. *Behav. Brain Res.,* 1981, 2, 189-209.

Dunnett, SB; Nathwani, F; Brasted, PJ. Medial prefrontal and neostriatal lesions disrupt performance in an operant delayed alternation task in rats. *Behav. Brain Res.,* 1999, 106, 13-28.

Dowie, MJ; Howard, ML; Nicholson, LF; Faull, RL; Hannan, AJ; Glass, M. Behavioural and molecular consequences of chronic cannabinoid treatment in Huntington's disease transgenic mice. *Neuroscience,* 2010, 170, 324-36.

Emerich, DF; Cain, CK; Greco, C; Saydoff, JA; Hu, ZY; Liu, H; Lindner, MD. Cellular delivery of human CNTF prevents motor and cognitive dysfunction in a rodent model of Huntington's disease. *Cell. Transplant.,* 1997, 6, 249-66.

Ennaceur, A. Effects of lesions of the substantia innominata/ventral Pallidum, globus pallidus and medial septum on rat's performance in object-recognition and radial-maze tasks: physostigmine and amphetamine treatments. *Pharmacol. Res.,* 1998, 38, 251-63.

Fernández-Ruiz, J; Doudet, D; Aigner, TG. Long-term cognitive impairment in MPTP-treated rhesus monkeys. *NeuroReport,* 1995, 7, 102-4.

Fernández-Ruiz, J; Doudet, D; Aigner, TG. Spatial memory improvement by levodopa in parkinsonian MPTP-treated monkeys. *Psychopharmacology,* 1999, 147, 104-7.

Ferro, MM; Bellissimo, MI; Anselmo-Franci, JA; Angellucci, ME; Canteras, NS; Da Cunha, C. Comparison of bilaterally 6-OHDA- and MPTP-lesioned rats as models of the early phase of Parkinson's disease: histological, neurochemical, motor and memory alterations. *J. Neurosci. Methods,* 2005, 148, 78-87.

Fielding, SA; Brooks, SP; Klein, A; Bayram-Weston, Z; Jones, L; Dunnett, SB. Profiles of motor and cognitive impairment in the transgenic rat model of Huntington's disease. *Brain Res. Bull.,* 2012, 88, 223-36.

Fleming, SM; Fernagut, PO; Chesselet, MF. Genetic mouse models of parkinsonism: strengths and limitations. *NeuroRx,* 2005, 2, 495-503.

Floresco, SB; Seamans, JK; Phillips, AG. Differential effects of lidocaine infusions into the ventral CA1/subiculum or the nucleus accumbens on the acquisition and retention of spatial information. *Behav. Brain Res.*, 1996, 81, 163-71.

Freichel C; Neumann M; Ballard T; Müller V; Woolley M; Ozmen L; Borroni E; Kretzschmar HA; Haass C; Spooren W; Kahle PJ. Age-dependent cognitive decline and amygdala pathology in alpha-synuclein transgenic mice. *Neurobiol. Aging,* 2007, 28, 1421-35.

Furtado, JC; Mazurek, MF. Behavioral characterization of quinolinate-induced lesions of the medial striatum: relevance for Huntington's disease. *Exp. Neurol.,* 1996, 138, 158-68.

Galey, D; Durkin, T; Sifakis, G; Kempf, E; Jaffard, R. Facilitation of spontaneous and learned spatial behaviours following 6-hydroxydopamine lesions of the lateral septum: a cholinergic hypothesis. *Brain Res.,* 1985, 340, 171-4.

Gal, G; Joel, D; Gusak, O; Feldon, J; Weiner, I. The effects of electrolytic lesion to the shell subterritory of the nucleus accumbens on delayed non-matching-to-sample and four-arm baited eight-arm radial-maze tasks. *Behav. Neurosci.,* 1997, 111, 92-103.

Garside, S; Furtado, JC; Mazurek, MF. Dopamine-glutamate interactions in the striatum: behaviourally relevant modification of excitotoxicity by dopamine receptor-mediated mechanisms. *Neuroscience,* 1996, 75, 1065-74.

Gasbarri, A; Sulli, A; Innocenzi, R; Pacitti, C,; Brioni, JD. Spatial memory impairment induced by lesion of the mesohippocampal dopaminergic system in the rat. *Neuroscience,* 1996, 74, 1037-44.

Giralt, A; Puigdellívol, M; Carretón, O; Paoletti, P; Valero, J; Parra-Damas, A; Saura, CA; Alberch, J; Ginés, S. Long-term memory deficits in Huntington's disease are associated with reduced CBP histone acetylase activity. *Hum. Mol. Genet.,* 2012, 21, 1203-16.

Gross, CG; Chorover, SL; Cohen, SM. Caudate, cortical, hippocampal and dorsal thalamic lesions in rats: Alternation and Hebb-Williams maze performance. *Neuropsychologia,* 1965, 3, 53-68.

Hagan, JJ; Alpert JE; Morris RG; Iversen, SD. The effects of central catecholamine depletions on spatial learning in rats. *Behav. Brain Res.*, 1983, 9, 83-104.

Haik, KL; Shear, DA; Hargrove, C; Patton, J; Mazei-Robison, M; Sandstrom, MI; Dunbar, GL. 7-nitroindazole attenuates 6-hydroxydopamine-induced spatial learning deficits and dopamine neuron loss in a presymptomatic animal model of Parkinson's disease. *Exp. Clin. Psychopharmacol.,* 2008, 16, 178-89.

Haik, KL; Shear, DA; Schroeder, U; Sabel, BA; Dunbar, GL. Quinolinic acid released from polymeric brain implants causes behavioral and neuroanatomical alterations in a rodent model of Huntington's disease. *Exp. Neurol.,* 2000, 163, 430-9.

Hefco, V; Yamada, K; Hefco, A; Hritcu, L; Tiron, A; Nabeshima, T. Role of the mesotelencephalic dopamine system in learning and memory processes in the rat. *Eur. J. Pharmacol.,* 2003, 475, 55-60.

Heng, MY; Detloff, PJ; Albin, RL. Rodent genetic models of Huntington disease. *Neurobiol. Dis.,* 2008, 32, 1-9.

Ho, YJ; Ho, SC; Pawlak, CR; Yeh, KY. Effects of D-cycloserine on MPTP-induced behavioral and neurological changes: potential for treatment of Parkinson's disease dementia. *Behav. Brain Res.,* 2011, 219, 280-90.

Holzmann, C; Dräger, D; Mix, E; Hawlitschka, A; Antipova, V; Benecke, R; Wree, A. Effects of intrastriatal botulinum neurotoxin A on the behavior of Wistar rats. *Behav. Brain Res.,* 2012, 234, 107-16.

Hsieh, MH; Ho, SC; Yeh, KY; Pawlak, CR; Chang, HM; Ho, YJ,; Lai, TJ; Wu, FY. Blockade of metabotropic glutamate receptors inhibits cognition and neurodegeneration in an MPTP-induced Parkinson's disease rat model. *Pharmacol. Biochem. Behav.,* 2012, 102, 64-71.

Itier, JM; Ibanez, P; Mena, MA; Abbas, N; Cohen-Salmon, C; Bohme, GA; Laville, M; Pratt, J; Corti, O; Pradier, L; Ret, G; Joubert, C; Periquet, M; Araujo, F; Negroni, J; Casarejos, MJ; Canals, S; Solano, R; Serrano, A; Gallego, E; Sanchez, M; Denefle, P; Benavides, J; Tremp, G; Rooney, TA; Brice, A; Garcia de Yebenes, J. Parkin gene inactivation alters behaviour and dopamine neurotransmission in the mouse. *Hum. Mol. Genet.,* 2003, 12, 2277-91.

Jeljeli, M; Strazielle, C; Caston, J; Lalonde, R. Effects of electrolytic lesions of the lateral pallidum on motor coordination, spatial learning, and regional brain variations of cytochrome oxidase activity in rats. *Behav. Brain Res.,* 1999, 102, 61-71.

Joel, D; Ayalon, L; Tarrasch, R; Veenman, L; Feldon, J; Weiner, I. Electrolytic lesion of globus pallidus ameliorates the behavioral and neurodegenerative effects of quinolinic acid lesion of the striatum: a potential novel treatment in a rat model of Huntington's disease. *Brain Res.,* 1998, 787, 143-8.

Kahle, PJ; Neumann, M; Ozmen, L; Muller, V; Jacobsen, H; Schindzielorz, A; Okochi, M; Leimer, U; van Der Putten, H; Probst, A; Kremmer, E; Kretzschmar, HA; Haass, C. Subcellular localization of wild-type and Parkinson's disease-associated mutant alpha - synuclein in human and transgenic mouse brain. *J. Neurosci.,* 2000, 20, 6365-73.

Kessler, J; Markowitsch, HJ. Delayed-alternation performance after kainic acid lesions of the thalamic mediodorsal nucleus and the ventral tegmental area in the rat. *Behav. Brain Res.,* 1981, 3, 125-30.

Kirch, RD; Meyer, PT; Geisler, S; Braun, F; Gehrig, S; Langen, KJ; von Hörsten, S; Nikkhah, G; Cassel, JC; Döbrössy, MD. Early deficits in declarative and procedural memory dependent behavioral function in a transgenic rat model of Huntington's disease. *Behav. Brain Res.,* 2013, 239, 15-26.

Kolb, B. Studies on the caudate-putamen and the dorsomedial thalamic nucleus of the rat: implications for mammalian frontal-lobe functions. *Physiol. Behav.,* 1977, 18, 237-44.

Kopalli, SR; Noh, SJ; Koppula, S; Suh, YH. Methylparaben protects 6-hydroxydopamine-induced neurotoxicity in SH-SY5Y cells and improved behavioral impairments in mouse model of Parkinson's disease. *Neurotoxicology,* 2013, 34, 25-32.

Kumar, P; Kaundal, RK; More, S; Sharma, SS. Beneficial effects of pioglitazone on cognitive impairment in MPTP model of Parkinson's disease. *Behav. Brain Res.,* 2009, 197, 398-403.

Lescaudron, L; Unni, D; Dunbar GL. Autologous adult bone marrow stem cell transplantation in an animal model of huntington's disease: behavioral and morphological outcomes. *Int. J. Neurosci.,* 2003, 113, 945-56.

Levin, ED; Galen, DM; Ellison, GD. Chronic haloperidol effects on oral movements and radial-arm maze performance in rats. *Pharmacol. Biochem. Behav.,* 1987, 26, 1-6.

Lex, B; Sommer, S; Hauber, W. The role of dopamine in the dorsomedial striatum in place and response learning. *Neuroscience,* 2011, 172, 212-8.

Li, X; Redus, L; Chen, C; Martinez, PA; Strong, R; Li, S; O'Connor, JC. Cognitive dysfunction precedes the onset of motor symptoms in the MitoPark mouse model of Parkinson's disease. *PLoS One,* 2013, 8(8), e71341.

Lione, LA; Carter, RJ; Hunt, MJ; Bates, GP; Morton, AJ; Dunnett, SB. Selective discrimination learning impairments in mice expressing the human Huntington's disease mutation. *J. Neurosci.,* 1999, 19, 10428-37.

Luchtman, DW; Meng, Q; Song, C. Ethyl-eicosapentaenoate (E-EPA) attenuates motor impairments and inflammation in the MPTP-probenecid mouse model of Parkinson's disease. *Behav. Brain Res.,* 2012, 226, 386-96.

Lüesse, HG; Schiefer, J; Spruenken, A; Puls, C; Block, F; Kosinski, CM. Evaluation of R6/2 HD transgenic mice for therapeutic studies in Huntington's disease; behavioral testing and impact of diabetes mellitus. *Behav. Brain Res.,* 2001, 126, 185-95.

Magen, I; Fleming, SM; Zhu, C; Garcia, EC; Cardiff, KM; Dinh, D; De La Rosa, K; Sanchez, M; Torres, ER; Masliah, E; Jentsch, JD; Chesselet, MF. Cognitive deficits in a mouse model of pre-manifest Parkinson's disease. *Eur. J. Neurosci.,* 2012, 35, 870-8.

Mangiarini, L; Sathasivam, K; Seller, M; Cozens, B; Harper, A; Hetherington, C; Lawton, M; Trottier, Y; Lehrach, H; Davies, SW; Bates, GP. Exon 1 of the HD gene with an expanded CAG repeat is sufficient to cause a progressive neurological phenotype in transgenic mice. *Cell,* 1996, 87, 493-506.

Martel, G; Blanchard, J; Mons, N; Gastambide, F; Micheau, J; Guillou, JL. Dynamic interplays between memory systems depend on practice: the hippocampus is not always the first to provide solution. *Neuroscience,* 2007, 150, 743-53.

Mitchell, JA; Hall, G. Caudate-putamen lesions in the rat may impair or potentiate maze learning depending upon availability of stimulus cues and relevance of response cues. *Q. J. Exp. Psychol. B.,* 1988, 40, 243-58.

Mitchell, JA; Channell, S; Hall, G. Response-reinforcer associations after caudate-putamen lesions in the rat: spatial discrimination and overshadowing-potentiation effects in instrumental learning. *Behav. Neurosci.,* 1985, 99, 1074-88.

Miyoshi E; Wietzikoski S; Camplessei M; Silveira R; Takahashi RN; Da Cunha C. Impaired learning in a spatial working memory version and in a cued version of the water maze in rats with MPTP-induced mesencephalic dopaminergic lesions. *Brain Res. Bull.,* 2002, 58, 41-7.

Mogensen, J; Iversen, IH; Divac, I. Neostriatal lesions impaired rats' delayed alternation performance in a T-maze but not in a two-key operant chamber. *Acta Neurobiol. Exp.,* 1987, 47, 45-54.

Mura, A; Feldon, J. Spatial learning in rats is impaired after degeneration of the nigrostriatal dopaminergic system. *Mov. Disord.,* 2003, 18, 860-71.

Murphy, KP; Carter, RJ; Lione, LA; Mangiarini, L; Mahal, A; Bates, GP; Dunnett, SB; Morton, AJ. Abnormal synaptic plasticity and impaired spatial cognition in mice transgenic for exon 1 of the human Huntington's disease mutation. *J. Neurosci.,* 2000 , 20, 5115-23.

Nguyen, HP; Kobbe, P; Rahne, H; Wörpel, T; Jäger, B; Stephan, M; Pabst, R; Holzmann, C; Riess, O; Korr, H; Kántor, O; Petrasch-Parwez, E; Wetzel, R; Osmand, A; von Hörsten, S. Behavioral abnormalities precede neuropathological markers in rats transgenic for Huntington's disease. *Hum. Mol. Genet.,* 2006, 15, 3177-94.

Nithianantharajah, J; Barkus, C; Murphy, M; Hannan, AJ. Gene-environment interactions modulating cognitive function and molecular correlates of synaptic plasticity in Huntington's disease transgenic mice. *Neurobiol. Dis.,* 2008, 29, 490-504.

Nuber, S; Petrasch-Parwez, E; Winner, B; Winkler, J; von Hörsten, S; Schmidt, T; Boy, J; Kuhn, M; Nguyen, HP; Teismann, P; Schulz, JB; Neumann, M; Pichler, BJ; Reischl, G; Holzmann, C; Schmitt, I; Bornemann, A; Kuhn, W; Zimmermann, F; Servadio, A; Riess, O. Neurodegeneration and motor dysfunction in a conditional model of Parkinson's disease. *J. Neurosci.,* 2008, 28, 2471-84.

Olmstead, CE; Villablanca, JR; Marcus, RJ; Avery, DL. Effects of caudate nuclei or frontal cortex ablations in cats. IV. Bar pressing, maze learning, and performance. *Exp. Neurol.,* 1976, 53, 670-93.

Packard, MG; Hirsh, R; White, NM. Differential effects of fornix and caudate nucleus lesions on two radial maze tasks: evidence for multiple memory systems. *J. Neurosci.,* 1989, 9, 1465-72.

Paumier, KL; Sukoff Rizzo, SJ; Berger, Z; Chen, Y; Gonzales, C; Kaftan, E; Li, L; Lotarski, S; Monaghan, M; Shen, W; Stolyar, P; Vasilyev, D; Zaleska, M; Hirst, DW; Dunlop, J. Behavioral characterization of A53T mice reveals early and late stage deficits related to Parkinson's disease. *PLoS One,* 2013, 8(8), e70274.

Perez, FA; Palmiter, RD. Parkin-deficient mice are not a robust model of parkinsonism. *Proc. Natl. Acad. Sci. USA,* 2005, 102, 2174-9.

Perry, JC; Da Cunha, C; Anselmo-Franci, J; Andreatini, R; Miyoshi, E; Tufik, S; Vital, MA. Behavioural and neurochemical effects of phosphatidylserine in MPTP lesion of the substantia nigra of rats. *Eur. J. Pharmacol.,* 2004, 484, 225-33.

Pioli, EY; Meissner, W; Sohr, R; Gross, CE; Bezard, E; Bioulac, BH. Differential behavioral effects of partial bilateral lesions of ventral tegmental area or substantia nigra pars compacta in rats. *Neuroscience,* 2008, 153, 1213-24.

Prediger, RD; Batista, LC; Medeiros, R; Pandolfo, P; Florio, JC; Takahashi, RN. The risk is in the air: Intranasal administration of MPTP to rats reproducing clinical features of Parkinson's disease. *Exp. Neurol.,* 2006, 202, 391-403.

Pisa, M; Cyr, J. Regionally selective roles of the rat's striatum in modality-specific discrimination learning and forelimb reaching. *Behav. Brain Res.,* 1990, 37, 281-92.

Pistell, PJ; Nelson, CM; Miller, MG, Spangler, EL; Ingram, DK; Devan, BD. Striatal lesions interfere with acquisition of a complex maze task in rats. *Behav. Brain Res.,* 2009, 197, 138-4.

Potegal, M. Role of the caudate nucleus in spatial orientation of rats. *J. Comp. Physiol. Psychol.,* 1969, 69, 756-64.

Ramos-Moreno, T; Castillo, CG; Martínez-Serrano, A. Long term behavioral effects of functional dopaminergic neurons generated from human neural stem cells in the rat 6-OH-DA Parkinson's disease model. Effects of the forced expression of BCL-X(L). *Behav. Brain Res.,* 2012, 232, 225-32.

Reksidler, AB; Lima, MM; Zanata, SM; Machado, HB; da Cunha, C; Andreatini, R; Tufik S, Vital MA. The COX-2 inhibitor parecoxib produces neuroprotective effects in MPTP-lesioned rats. *Eur. J. Pharmacol.,* 2007, 560, 163-75.

Rex, A; Fink, H. Cholecystokinin tetrapeptide improves water maze performance of neonatally 6-hydroxydopamine-lesioned young rats. *Pharmacol. Biochem. Behav.,* 2004, 79, 109-17.

Rieker, C; Dev, KK; Lehnhoff, K; Barbieri, S; Ksiazek, I; Kauffmann, S; Danner, S; Schell, H; Boden, C; Ruegg, MA; Kahle, PJ; van der Putten, H; Shimshek, DR. Neuropathology in mice expressing mouse alpha-synuclein. *PLoS One,* 2011, 6(9), e24834.

Roberts, AC; De Salvia, MA; Wilkinson, LS; Collins, P; Muir, JL; Everitt, BJ; Robbins, TW. 6-Hydroxydopamine lesions of the prefrontal cortex in monkeys enhance performance on an analog of the Wisconsin Card Sort Test: possible interactions with subcortical dopamine. *J. Neurosci.,* 1994, 14, 2531-44.

Sagratella, S; Scotti de Carolis, A; Diana, G; Domenici, MR; Popoli, P. Selective reduction of hippocampal dentate frequency-potentiation in striatally lesioned rats with impaired place learning. *Brain Res.,* 1994, 660, 66-72.

Schacter, GB; Yang, CR; Innis, NK; Mogenson, GJ. The role of the hippocampal-nucleus accumbens pathway in radial-arm maze performance. *Brain Res.,* 1989, 494, 339-49.

Schwartzbaum, JS; Donovick, PJ. Discrimination reversal and spatial alternation associated with septal and caudate dysfunction in rats. *J. Comp. Physiol. Psychol.,* 1968, 65, 83-92.

Shear, DA; Dong, J; Gundy, CD; Haik-Creguer, KL; Dunbar, GL. Comparison of intrastriatal injections of quinolinic acid and 3-nitropropionic acid for use in animal models of Huntington's disease. *Prog. Neuropsychopharmacol. Biol. Psychiatry,* 1998, 22, 1217-40.

Shear, DA; Dong, J; Haik-Creguer, KL; Bazzett, TJ; Albin, RL; Dunbar, GL. Chronic administration of quinolinic acid in the rat striatum causes spatial learning deficits in a radial arm water maze task. *Exp. Neurol.,* 1998, 150, 305-11.

Simon, H; Taghzouti, K; Le Moal, M. Deficits in spatial-memory tasks following lesions of septal dopaminergic terminals in the rat. *Behav Brain Res,* 1986, 19, 7-16.

Slow, EJ; van Raamsdonk, J; Rogers, D; Coleman, SH; Graham, RK; Deng, Y; Oh, R; Bissada, N; Hossain, SM; Yang, YZ; Li, XJ; Simpson, EM; Gutekunst, CA; Leavitt, BR; Hayden, MR. Selective striatal neuronal loss in a YAC128 mouse model of Huntington disease. *Hum. Mol. Genet.,* 2003, 12, 1555-67.

Stam, CJ; de Bruin, JP; van Haelst, AM; van der Gugten, J; Kalsbeek, A. Influence of the mesocortical dopaminergic system on activity, food hoarding, social-agonistic behavior, and spatial delayed alternation in male rats. *Behav. Neurosci.,* 1989, 103, 24-35.

Sutherland, RJ; Rodriguez, AJ. The role of the fornix/fimbria and some related subcortical structures in place learning and memory. *Behav. Brain Res.,* 1989, 32, 265-77.

Tabrizi, SJ; Workman, J; Hart, PE; Mangiarini, L; Mahal, A; Bates, G; Cooper, JM; Schapira, AHV. Mitochondrial dysfunction and free radical damage in the Huntington R6/2 transgenic mouse. *Ann. Neurol.,* 2000, 47, 80-6.

Tadaiesky, MT; Dombrowski, PA; Figueiredo, CP; Cargnin-Ferreira, E; Da Cunha, C; Takahashi, RN. Emotional, cognitive and neurochemical alterations in a premotor stage model of Parkinson's disease. *Neuroscience,* 2008, 156, 830-40.

Taghzouti, K; Louilot, A; Herman, JP; Le Moal, M; Simon, H. Alternation behavior, spatial discrimination, and reversal disturbances following 6-hydroxydopamine lesions in the nucleus accumbens of the rat. *Behav. Neural Biol.,* 1985, 44, 354-63.

Tanila, H; Björklund, M; Riekkinen, P Jr. Cognitive changes in mice following moderate MPTP exposure. *Brain Res. Bull.,* 1998, 45, 577-82.

Thifault, S; Krémarik, P; Lalonde, R. Effects of bilateral electrolytic lesions of the medial nucleus accumbens on exploration and spatial learning. *Arch. Physiol. Biochem.,* 1998, 106, 297-307.

Thompson, R; Yang, S. Retention of individual spatial reversal problems in rats with nigral, caudoputamenal, and reticular formation lesions. *Behav. Neural Biol.,* 1982, 34, 98-103.

van Raamsdonk, JM; Metzler, M; Slow, E; Pearson, J; Schwab, C; Carroll, J; Graham, RK; Leavitt, BR; Hayden, MR. Phenotypic abnormalities in the YAC128 mouse model of Huntington's disease are penetrant on multiple genetic backgrounds and modulated by strain. *Neurobiol. Dis.,* 2007, 26, 189-200.

von Hörsten, S; Schmitt, I; Nguyen, HP; Holzmann, C; Schmidt, T; Walther, T; Bader, M; Pabst, R; Kobbe, P; Krotova, J; Stiller, D; Kask, A; Vaarmann, A; Rathke-Hartlieb, S; Schulz, JB; Grasshoff, U; Bauer, I; Vieira-Saecker, AM; Paul, M; Jones, L; Lindenberg, KS; Landwehrmeyer, B; Bauer, A; Li, XJ; Riess, O. Transgenic rat model of Huntington's disease. *Hum. Mol. Genet.,* 2003, 12, 617-24.

Whishaw, IQ; Dunnett, SB. Dopamine depletion, stimulation or blockade in the rat disrupts spatial navigation and locomotion dependent upon beacon or distal cues. *Behav. Brain Res.,* 1985, 18, 11-29.

Whishaw, IQ; Mittleman, G; Bunch, T; Dunnett, SB. Impairments in the acquisition, retention and selection of spatial navigation strategies after medial caudate-putamen lesions in rats. *Behav. Brain Res.,* 1987, 24, 125-38.

Wikmark, RGE; Divac, I; Weiss, R. Retention of spatial delayed alternation in rats with lesions in the frontal lobes: implications for a comparative neuropsychology of the prefrontal system. *Brain Behav. Evol.,* 1973, 8, 329-39.

Wood, NI; Glynn, D; Morton, AJ. "Brain training" improves cognitive performance and survival in a transgenic mouse model of Huntington's disease. *Neurobiol. Dis.,* 2011, 42, 427-37.

Wood, NI; Carta, V; Milde, S; Skillings, EA; McAllister, CJ; Ang, YL; Duguid, A; Wijesuriya, N; Afzal, SM; Fernandes, JX; Leong, TW; Morton, AJ. Responses to environmental enrichment differ with sex and genotype in a transgenic mouse model of Huntington's disease. *PLoS One,* 2010, 5(2), e9077.

Zeef, DH; van Goethem, NP; Vlamings, R; Schaper, F; Jahanshahi, A; Hescham, S; von Hörsten, S; Prickaerts, J; Temel, Y. Memory deficits in the transgenic rat model of Huntington's disease. *Behav. Brain Res.,* 2012, 227, 194-8.

Zhou, W; Milder, JB; Freed, CR. Transgenic mice overexpressing tyrosine-to-cysteine mutant human alpha-synuclein: a progressive neurodegenerative model of diffuse Lewy body disease. *J. Biol. Chem.,* 2008, 283, 9863-70.

Zhu, XR; Maskri, L; Herold, C; Bader, V; Stichel, CC; Güntürkün, O; Lübbert, H. Non-motor behavioural impairments in parkin-deficient mice. *Eur. J. Neurosci.,* 2007, 26, 1902-11.

In: Spatial Orientation in Humans and Animals
Editor: Robert Lalonde
ISBN: 978-1-63463-001-6
© 2014 Nova Science Publishers, Inc.

Chapter 8

SPATIAL DEFICITS IN PATIENTS WITH CEREBELLAR DISORDERS

Robert Lalonde*
Université de Rouen, Dépt. Psychologie,
Mont-Saint-Aignan Cedex France

1. SPINOCEREBELLAR ATAXIA AND FRIEDREICH'S ATAXIA

Spinocerebellar ataxias (SCAs) comprise autosomal dominant diseases causing widespread gait (ataxia), disturbances in postural control, dysmetria (imprecise movements), and action tremor (Koeppen, 1998). The dysmetria usually occurs in the form of hypermetria (target overshooting), but hypometria may also occur, both related to the rate of rise of antagonist muscle activity (Manto et al., 1996, 1998). SCA1, SCA2, SCA3/Machado-Joseph (MJD), SCA5, and SCA7 diseases are caused by CAG (cytosine-adedine-guanine) triplet expansions leading to cerebellar and brainstem damage. In SCA3/MJD, there is additional damage to substantia nigra pars compacta, causing parkinsonian symptoms. SCA6 results in a highly specific late-onset cerebellar degeneration and SCA7 in additional damage to the retina. Over thirty SCAs have been identified with patterns of brain damage overlapping considerably at advanced stages, so that only a few exhibit behavioral symptoms and histological patterns distinct from each other (Seidel et al., 2012).

Among autosomal recessive forms of spinocerebellar atrophy, the most prevalent is Friedreich's ataxia, in which cerebellar atrophy is generally less extensive than in autosomal dominant SCAs (Campuzano et al., 1996; Corben et al., 2006, 2012). Instead, there is marked atrophy of spinocerebellar tracts and peripheral sensory nerves of dorsal and cranial root ganglia. Friedreich's ataxia is caused by a GAA triplet expansion that deletes the *FXN* gene situated on chromosome 9 and encoding frataxin, a protein involved in iron transport inside mitochondria (Campuzano et al., 1996; Delatycki et al., 1999). The mutation causes a redistribution of iron, copper, and zinc in the brain (Koeppen et al., 2012). The patients are susceptible to early death mostly from cardiac anomalies (Tsou et al., 2011) and prior to death

* Email: robert.lalonde@univ-rouen.fr

often display Cheyne-Stokes breathing associated with bulbar damage (Lalonde and Botez, 1991).

2. Spatial Short-Term and Working Memory

Spatial deficits in patients with heredodegenerative ataxias are summarized in Table 1. Patients with SCA3/MJD were impaired in forward and backward span in the block-tapping subtest of the Weschler adult intelligence scale-revised (WAIS-R) (Braga-Neto et al., 2012). Patients with SCA3/MJD disease were slower in visual information processing and had difficulty in shifting visual attention (Maruff et al., 1996), a function which may underlie the spatial deficits. Likewise, hereditary SCA patients of different subcategories were impaired in forward and backward span of the Corsi block-tapping test (Arroyo-Anllo and Botez-Marquard, 1998). On the contrary, block-tapping span was normal in a mixed group with cerebellar lesions, most with unilateral strokes (Schmahmann and Sherman, 1998). The block tapping subtest of the WAIS-R was also normal in a mixed group with strokes or with tumor or cyst resection (Ravizza et al., 2006). These results indicate that heredodegenerative ataxias with bilateral lesions may lead to more important deficits in short-term or working memory than unilateral ones.

Table 1. Spatial deficits in patients with heredodegenerative ataxias

Function	Tests	Reference
visuoconstructive skills	block design, object assembly	Botez-Marquard and Botez, 1993; Botez-Marquard et al., 1994; Kawai et al., 2004; Molinari et al., 2004; Schmahmann and Sherman, 1998
visuoconstructive skills	simple drawing	Arroyo-Anllo and Botez-Marquard, 1998
visuoconstructive skills	Rey complex figure	Botez-Marquard and Botez, 1993; Braga-Neto et al., 2012; Schmahmann and Sherman, 1998; Steinlin et al., 1999
visuospatial reasoning	Raven progressive matrices	Botez-Marquard and Botez, 1993; Kish et al., 1988; Orr et al., 1993
figure rotations	mental folding	Fehrenbach et al., 1984; Molinari et al., 2004
spatial learning	delayed alternation	El-Awar et al., 1991
working memory	block-tapping	Arroyo-Anllo and Botez-Marquard, 1998; Braga-Neto et al., 2012

3. Visuoconstructive Skills, Spatial Reasoning, and Spatial Long-Term Memory

Because dysmetria and action tremor are present, cognitive functions in patients with cerebellar disorders are best estimated by contrasting timed versus untimed versions of the block design subtest of the Weschler intelligence scale (Botez-Marquard and Botez, 1993). Likewise, copy of the Rey complex figure can be scored for accuracy independently of upper limb dysmetria (Schmahmann and Sherman, 1998). In a similar fashion, a spatial discrimination task can be conducted at stimulus presentation intervals shorter than those required for gaze shifting, thereby excluding the possibility that the deficit is caused by oculomotor disorders (Townsend et al., 1999).

3.1. Anti-Epileptic Intoxication

Two case reports of cerebellar atrophy caused by chronic phenytoin administration first indicated a role for the cerebellum in visuoconstructive skills and spatial reasoning (Botez et al., 1985). In both epileptic cases, cerebellar dysmetria and dysarthria improved after repeated oral administration of thiamine, a B-vitamin liable to be depleted in the central nervous system as a result of long-term phenytoin therapy (Botez and Young, 1991) and also in inherited degenerative ataxias (Pedraza and Botez, 1992). The first patient was evaluated with neuropsychological tests three times over a period of nearly two years. An amelioration in ataxic gait and dysarthria was paralleled by improved scores in block design and object assembly subtests of the Wechler intelligence scale as well as copy of the Rey complex figure. Moreover, a group of epileptic patients with radiologic evidence of cerebellar and brainstem damage caused by anti-epileptic intoxication had lower scores on block design and object assembly substests than those without such damage (Botez et al., 1989). Block design and object assembly scores were linearly correlated with cerebellar atrophy as determined by fourth ventricle ratios.

3.2. Heredodegenerative Ataxias

Hereditary SCAs cause relatively selective damage to the cerebellum and brainstem afferent regions, providing an appropriate sample population to test hypotheses first formulated by Botez et al. (1985) and Leiner et al. (1986) regarding the role of the cerebellum on cognitive processes. In comparison to healthy subjects, a group of hereditary SCA patients of different subcategories was impaired in timed and untimed versions of block design (Botez-Marquard and Botez, 1993), whereas patients with Friedriech's ataxia were only impaired in the timed version (Botez-Marquard and Botez, 1993; Fehrenbach et al., 1984; White et al., 2000). These results imply a specific deficit in visuoconstructive skills following severe as opposed to milder cerebellar atrophy. The deficit seen in patients with Friedriech's ataxia is more ambiguous, perhaps caused by mental slowing but also by a physical difficulty in manipulating the blocks. The untimed version of block design was also deficient in SCA3/MJD patients (Kawai et al., 2004). The timed version of block design was deficient in a separate hereditary SCA sample (Berent et al., 1990). In contrast to block design, both SCA and Friedreich's ataxia groups were impaired in copying accuracy of the Rey complex figure and in spatial reasoning of Raven standard progressive matrices, the latter comprising no movement on the part of the patient and thereby no possibility that dysmetria influenced the results (Botez-Marquard and Botez, 1993). Hereditary SCA patients were also inferior to controls in copying a simpler geometric design (Arroyo-Anllo and Botez-Marquard, 1998).

Copy of the Rey complex figure was also deficient in patients with SCA3/MJD (Braga-Neto et al., 2012) and nonprogressive cerebellar ataxia (Steinlin et al., 1999). The result in SCA3/MJD subjects was not reproduced in a subsequent study, nor was it seen specifically in those with autosomal dominant SCA1 or SCA2 (Bürk et al., 2003), probably due to the use of IQ-matched controls. Since cerebellar lesions can cause deficits in some IQ subtests (see above), this procedure amounts to testing a subcategory of these patient groups. The susceptibility of Raven standard progressive matrices to cerebellar damage was confirmed in patients with SCA1 (Kish et al., 1988; Orr et al., 1993). SCA3/MJD patients also scored lower than controls in figure reproduction of the Wechsler memory scale (Kawai et al., 2004). However, tests of visuospatial memory and attention were not impaired in patients with SCA1 patients (Bürk et al., 2001) or SCA6 (Globas et al., 2003) relative to IQ-matched controls.

A group with ideopathic cerebellar ataxia, presumably heredoataxic in nature, was deficient in object assembly and two tests of mental folding requiring no movement (differential aptitude test-spatial relations and Minnesota paper form board test) but not in judgment of Benton line orientation (Molinari et al., 2004), although normal subjects activate cerebellar subregions during judgment of Benton line orientation (Lee et al., 2005). Likewise, patients with Friedriech's ataxia were deficient in a mental folding test requiring the conversion of unfolded two-dimensional drawings into a three-dimensional one (Fehrenbach et al., 1984). Morover, hereditary SCA patients were impaired in the delayed alternation test (El-Awar et al., 1991).

The specific contribution of the cerebellum on the performance of block design, Rey complex figure, and Raven standard progressive matrices was further emphasized by the poorer performance seen in hereditary SCA patients with severe as opposed to milder cerebellar atrophy based on radiological evidence in the size of the fourth ventricle (Botez-Marquard et al., 1996). Block design and Raven standard progressive matrices were also lower in patients with more severe atrophy at the pontine level, a source of mossy fibers to the cerebellum.

3.3. Vascular or Mixed Groups

Patients with cerebellar lesions of vascular (ischemia or hemorrhage) origin have been evaluated in spatial tests along with those from mixed groups comprising vascular and other disorders. The vascular cases mostly comprise unilateral damage. In a mixed group (most with unilateral strokes but also with heredodegenerative ataxias, postinfectuous cerebellitis, or tumor removal), block design, object assembly, and copy of the Rey complex figure were impaired relative to control subjects (Schmahmann and Sherman, 1998). Likewise, object assembly and copy of the Rey complex figure were deficient in a single patient after unilateral cerebellar stroke (Botez-Marquard et al., 1994). In contrast, a group with left-sided cerebellar infarcts was deficient in time taken to draw the Rey complex figure but not in accuracy (Hokkanen et al., 2006). The spatial deficits reported by Schmahmann and Sherman (1998) were transient, probably because of their unilateral nature, as those seen in the patient with unilateral stroke (Botez-Marquard et al., 1994). The absence of any spatial deficit reported in patients with unilateral cerebellar stroke (infarcts or hemorrhages) (Beldarrain et al., 1997) is a further indication of the minimal long-term impact of unilateral cerebellar lesions.

Nevertheless, block design was deficient in a mixed group with cerebellar damage caused by hematoma, oedema, or tumor (Gottwald et al., 2004), as well as one with infarcts limited to the cerebellum or pontine nuclei (Malm et al., 1998) or the cerebellum alone (Neau et al., 2000), the vast majority in all three studies with unilateral lesions. On one hand, a mixed group with cerebellar damage caused by hematoma or tumor was impaired in figure memory subtest of the Wechsler memory scale but not block design or copy of the Rey complex figure (Gottwald et al., 2003). On the other, a mixed group comprising five patients with heredodegenerative ataxias and another with an ischemic lesion of the cerebellum with brainstem damage had lower scores than a control group in copying the Rey complex figure and in the Benton visual retention test (Daum et al., 1993). The recall of Benton figures was also impaired in a mixed cerebellar-lesioned group comprising heredodegenerative ataxias, tumor removal, or stroke (Bracke-Tolkmitt et al., 1989). In addition, a mixed group of patients with cerebellar lesions caused by stroke or tumor were impaired in a spatial discrimination task (Townsend et al., 1999). As seen in heredodegenerative ataxias (see above), the test of Raven progressive matrices was deficient in a spinocerebellar group of mixed etiologies (Hirono et al., 1991), as was a focal group with left-sided cerebellar ataxia in two tests of mental folding (Molinari et al., 2004). In a study comprising a mixed group with vascular lesions and tumor resection with no control group in the experimental design, individual cases were impaired in Raven standard progressive matrices and copy of the Rey complex figure (Baillieux et al., 2010).

3.4. Surgery

De Ribaupierre et al. (2008) studied patients diagnosed with a cerebellar lesion. Neuropsychological tests assessing short-term and anterograde memory, verbal and visuospatial modalities were performed pre-operatively. Memory deficits in at least one modality were identified in 81% of patients with cerebellar lesions.

4. Cerebellar-Related Pathways Underlying Spatial Performances

As noted in several reviews (Baillieux et al., 2008; Bellebaum and Daum, 2007; Daum and Ackermann, 1995, 1997; Daum et al., 1993; Justus and Ivry, 2001; Schmahmann, 1991; Timmann and Daum, 2007, 2010), the effects of cerebellar damage on spatial processing depend on disease etiologies and testing methods, but are certainly less severe than those observed after telencephalic lesions. The milder deficits confirm the hypothesis that these spatial deficits are caused by reduced metabolism or blood flow in brain regions connected with the cerebellum, in particular frontal and parietal association cortex as well as basal ganglia (Botez et al., 1991; Nguyen and Botez, 1998). Reduced metabolism in posterior parietal cortex may underlie deficits in spatial skills, while reduced metabolism in prefrontal cortex may impair planning an organized sequence of events, sometimes affected in cerebellar disease (Grafman et al., 1992).

In stroke patients, cerebellar lesions lead to contralateral loss of blood flow in neocortex, defined as cerebello-cortical diaschisis (Broich et al., 1987; Komaba et al., 2000; Sonmezoglu et al., 1993). This mechanism is explained by transsynaptic changes occurring in supratentorial areas as a secondary consequence of cerebellar atrophy. Without excluding the possible impact of mutated genes directly on the neocortex in cases of heredodegenerative ataxias, Botez et al. (1985, 1991) first proposed that cognitive deficits associated with cerebellar syndromes are the result of diaschisis in efferent structures such as basal ganglia and neocortex or afferent structures such as basal pontine nuclei. In particular, cerebello-cerebral diaschisis may be ascribed to disruption of anterograde cerebello-thalamo-cortical or cerebello-midbrain-striatal projections, the latter implying ascending dopaminergic neurons to striatum and prefrontal cortex (chapter 3). Reduced regional brain metabolism or blood flow of supratentorial areas may also be caused by disruption of the retrograde pathway from the cerebellum to association cortices via basal pontine nuclei. In patients with crossed cerebellar hypoperfusion, cerebellar metabolism was linearly correlated with pontine metabolism in the opposite hemisphere (Yamaguchi et al., 1992). Moreover, patients with pontine infarcts displayed contralateral hypoperfusion of the cerebellum along with ipsilateral hypoperfusion of the frontoparietal cortex (Fazekas et al., 1993). Likewise, diffuse magnetic resonance imaging combined with white matter tractography indicates disrupted cerebello-cortical connections in Friedreich's ataxia (Zalesky et al., 2014). These data indicate that interruption of the pontocerebellar tract may be responsible for the neocortical hypometabolism seen after cerebellar lesions and likely for the neuropsychological deficits as well. It remains to be determined whether regional brain metabolism is correlated with cognitive deficits in cerebellar-damaged patients in the same fashion as has already been demonstrated with respect to motor symptoms (Kluin et al., 1988; Rosenthal et al., 1988).

References

Arroyo-Anllo, EM; Botez-Marquard, T. Neurobehavioral dimensions of olivopontocerebellar atrophy. *J. Clin. Exp. Neuropsychol.,* 1998, 20, 52-9.

Baillieux, H; De Smet, HJ; Paquier, PF; De Deyn, PP; Mariën, P. Cerebellar neurocognition: insights into the bottom of the brain. *Clin. Neurol. Neurosurg.,* 2008, 110, 763-73.

Baillieux, H; De Smet, HJ; Dobbeleir, A; Paquier, PF; De Deyn, PP; Mariën, P. Cognitive and affective disturbances following focal cerebellar damage in adults: a neuropsychological and SPECT study. *Cortex,* 2010, 46, 869-79.

Beldarrain, MG; Garcia-Manco, J-C; Quintana, JM; Rodeno, E. Diaschisis and neuropsychological performance after cerebellar stroke. *Eur. Neurol.,* 1997, 7, 82-9.

Bellebaum, C; Daum, I. Cerebellar involvement in executive control. *Cerebellum,* 2007, 6, 184-92.

Berent, S; Giordani, B; Gilman, S; Junck, L; Lehtinen, S; Markel, DS; Boivin, M; Kluin, KJ; Parks, R; Koeppe, RA. Neuropsychological changes in olivopontocerebellar atrophy. *Arch. Neurol.,* 1990, 47, 997-1001.

Botez, MI; Young, SN. Effects of anticonvulsant treatment and low levels of folate and thiamine on amine metabolites in cerebrospinal fluid. *Brain,* 1991, 114, 333-48.

Botez, MI; Botez, T; Elie, R; Attig, E. Role of the cerebellum in complex human behavior. *Ital. J. Neurol. Sci.,* 1989, 10, 291-300.

Botez, MI; Gravel, J; Attig, E; Vézina, 1985. Reversible chronic cerebellar atxia after phenytoin intoxication: possible role of cerebellum in cognitive thought. *Neurology,* 1985, 35, 1152-7.

Botez, MI; Léveillé, J; Lambert, R; Botez, T. Single photon emission computed tomography (SPECT) in cerebellar disease: cerebello-cerebral diaschisis. *Eur. Neurol.,* 1991, 31, 405-12.

Botez-Marquard, T; Botez, MI. Cognitive behavior in heredodegenerative ataxias. *Eur. Neurol.,* 1993, 33, 351-7.

Botez-Marquard, T; Léveillé, J; Botez, MI. Neuropsychological functioning in unilateral cerebellar damage. *Can. J. Neurol. Sci.,* 1994, 21, 353-7.

Botez-Marquard, T; Pedraza, OL; Botez, MI. Neuroradiological correlates of neuropsychological disorders in olivopontocerebellar atrophy (OPCA). *Eur. J. Neurol.,* 1996, 3, 89-97.

Bracke-Tolkmitt, R; Linden, A; Canavan, AGM; Rockstroh, B; Scholz, E; Wessel, K; Diener, H-C. The cerebellum contributes to mental skills. *Behav. Neurosci.,* 1989, 103, 442-6.

Braga-Neto, P; Pedroso, JL; Alessi, H; Dutra, LA; Felício, AC; Minett, T; Weisman, P; Santos-Galduroz, RF; Bertolucci, PH; Gabbai, AA; Barsottini, OG. Cerebellar cognitive affective syndrome in Machado Joseph disease: core clinical features. *Cerebellum,* 2012, 11, 549-56.

Broich, K; Hartmann, A; Biersack, H-J; Horn, R. Crossed cerebello-cerebral diaschisis in a patient with cerebellar infarction. *Neurosci. Lett.,* 1987, 83, 7-12.

Bürk, K; Bösch, S; Globas, C; Zühlke, C; Daum, I; Klockgether, T; Dichgans, J. Executive dysfunction in spinocerebellar ataxia type 1. *Eur. Neurol.,* 2001, 46, 43-8.

Bürk, K; Globas, C; Bösch, S; Klockgether, T; Zühlke, C; Daum, I; Dichgans, J. Cognitive deficits in spinocerebellar ataxia type 1, 2, and 3. *J. Neurol.,* 2003, 250, 207-11.

Campuzano, V; Montermini, L; Moltò, MD; Pianese, L; Cossée, M; Cavalcanti, F; Monros, E; Rodius, F; Duclos, F; Monticelli, A; Zara, F; Cañizares, J; Koutnikova, H; Bidichandani, SI; Gellera, C; Brice, A; Trouillas, P; De Michele, G; Filla, A; De Frutos, R; Palau, F; Patel, PI; Di Donato, S; Mandel, JL; Cocozza, S; Koenig, M; Pandolfo, M. Friedreich's ataxia: autosomal recessive disease caused by an intronic GAA triplet repeat expansion. *Science,* 1996, 271, 1423-27.

Corben, LA; Georgiou-Karistianis, N; Bradshaw, JL; Evans-Galea, MV; Churchyard, AJ; Delatycki, MB. Characterising the neuropathology and neurobehavioural phenotype in Friedreich ataxia: a systematic review. *Adv. Exp. Med. Biol.,* 2012, 769, 169-84.

Corben, LA; Georgiou-Karistianis, N; Fahey, MC; Storey, E; Churchyard, A; Horne, M; Bradshaw, JL; Delatycki, MB. Towards an understanding of cognitive function in Friedreich ataxia. *Brain Res. Bull.,* 2006, 70, 197-202.

Daum, I; Ackermann, H. Cerebellar contributions to cognition. *Behav. Brain Res.,* 1995, 67, 201-20.

Daum, I; Ackermann, H. Neuropsychological abnormalities in cerebellar syndromes- fact or fiction? *Int. Rev. Neurobiol.,* 1997, 41, 455-71.

Daum, I; Ackermann, H; Schugens, MM; Reimold, C; Dichgans, J; Birbaumer, N. The cerebellum and cognitive functions in humans. *Behav. Neurosci.,* 1993, 107, 411-9.

Delatycki, MB; Camakaris, J; Brooks, H; Evans-Whipp, T; Thorburn, DR; Williamson, R; Forrest, SM. Direct evidence that mitochondrial iron accumulation occurs in Friedreich ataxia. *Ann. Neurol.*, 1999, 45, 673-5.

de Ribaupierre, S; Ryser, C; Villemure, JG; Clarke, S. Cerebellar lesions: is there a lateralisation effect on memory deficits? *Acta Neurochir.,* 2008, 150, 545-50.

El-Awar, M; Kish, S; Oscar-Berman, M; Robitaille, Y; Schut, L; Freedman, M. Selective delayed alternation deficits in dominantly inherited olivopontocerebellar atrophy. *Brain Cogn.,* 1991, 16, 121-9.

Fazekas, F; Payer, F; Valetitsch, H; Schmidt, R; Flooh, E. Brain stem infarction and diaschisis: a SPECT perfusion study. *Stroke,* 1993, 24, 1162-6.

Fehrenbach, RA; Wallesch, C-W; Claus, D. Neuropsychologic findings in Friedreich's ataxia. *Arch. Neurol.*, 1984, 41, 306-8.

Globas, C; Bösch, S; Zühlke, C; Daum, I; Dichgans, J; Bürk, K. The cerebellum and cognition. Intellectual function in spinocerebellar ataxia type 6 (SCA6). *J. Neurol.,* 2003, 250, 1482-7.

Gottwald, B; Mihajlovic, Z; Wilde, B; Mehdorn, HM. Does the cerebellum contribute to specific aspects of attention? *Neuropsychologia,* 2003, 41, 1452-60.

Gottwald, B; Wilde, B; Mihajlovic, Z; Mehdorn, HM. Evidence for distinct cognitive deficits after focal cerebellar lesions. *J. Neurol. Neurosurg. Psychiatry,* 2004, 75, 1524-31.

Grafman, J; Litvan, I; Massaquoi, S: Stewart, M; Sirigu, A; Hallett, M. Cognitive planning deficit in patients with cerebellar atrophy. *Neurology,* 1992, 42, 1493-6.

Hirono, N; Yamadori, A; Kameyama, M; Mezaki, T; Abe, K. Spinocerebellar degeneration (SCD): cognitive disturbances. *Acta Neurol. Scand.*, 1991, 84, 226-30.

Hokkanen, LS; Kauranen, V; Roine, RO; Salonen, O; Kotila, M. Subtle cognitive deficits after cerebellar infarcts. *Eur. J. Neurol.,* 2006, 13, 161-70.

Justus, TC; Ivry, RB. The cognitive neuropsychology of the cerebellum. *Int. Rev. Neuropsychiatry,* 2001, 13, 276-82.

Kawai, Y; Takeda, A; Abe, Y; Washimi, Y; Tanaka, F; Sobue, G. Cognitive impairments in Machado-Joseph disease. *Arch. Neurol.,* 2004, 61, 1757-60.

Kish, SJ; El-Awar, M; Schut, L; Leach, L; Oscar-Berman, M; Freedman, M. Cognitive deficits in olivopontocerebellar atrophy: implications for the cholinergic hypothesis of Alzheimer's dementia. *Ann. Neurol.,* 1988, 24, 200-6.

Kluin, KK; Gilman, S; Markel, DS; Koeppe, RA; Rosenthal, G; Junck, L. Speech disorders in olivopontocerebellar atrophy correlate with positron emission tomography findings. *Ann. Neurol.,* 1988, 23, 547-54.

Koeppen, AH. The hereditary ataxias. *J. Neuropathol. Exp. Neurol.*, 1998, 57, 531-43.

Koeppen, AH; Ramirez, RL; Yu, D; Collins, SE; Qian, J; Parsons, PJ; Yang, KX; Chen, Z; Mazurkiewicz, JE; Feustel, PJ. Friedreich's ataxia causes redistribution of iron, copper, and zinc in the dentate nucleus. *Cerebellum,* 2012, 11, 845-60.

Komaba, Y; Osono, E; Kitamura, S; Kayatama, Y. Crossed cerebellocerebral diaschisis in patients with cerebellar stroke. *Acta Neurol. Scand.*, 2000, 101, 8-12.

Lalonde, R; Botez, MI. Death from bulbar involvement in Friedreich's ataxia. *Med. Hypoth.,* 1991, 36, 250-2.

Lee, TM; Liu, HL; Hung, KN; Pu, J; Ng, YB; Mak, AK; Gao, JH; Chan, CC. The cerebellum's involvement in the judgment of spatial orientation: a functional magnetic resonance imaging study. *Neuropsychologia,* 2005, 43, 1870-7.

Leiner, HC; Leiner, AL; Dow, RS. Does the cerebellum contribute to mental skills? *Behav. Neurosci.,* 1986, 100, 443-54.

Malm, J; Kristensen, B; Karlsson, T; Carlberg, B; Fagerlund, M; Olsson, T Cognitive impairment in young adults with infratentorial infarcts. *Neurology,* 1998, 51, 433-40.

Maruff, P; Tyler, P; Burt, T; Currie, B; Burns, C; Currie, J. Cognitive deficits in Machado-Joseph disease. *Ann. Neurol.,* 1996, 40, 421-7.

Manto, M; Godaux, E; Jacquy, J; Hildebrand, J. Cerebellar hypermetria associated with a selective decrease in the rate of rise of antagonist activity. *Ann. Neurol.,* 1996, 39, 271-4.

Manto, M-U; Setta, F; Jacquy, J; Godaux, E; Hildebrand, J; Roland, H; Blum, S; Brohee, P. Different types of cerebellar hypometria associated with a disctinct topography of the lesion in cerebellum. *J. Neurol. Sci.,* 1998, 158, 88-95.

Molinari, M; Petrosini, L; Misciagna, S; Leggio, MG. Visuospatial abilities in cerebellar disorders. *J. Neurol. Neurosurg. Psychiatry,* 2004, 75, 235-40.

Neau, JP; Arroyo-Anllo, E; Bonnaud, V; Ingrand, P; Gil, R. Neuropsychological disturbances in cerebellar infarcts. *Acta Neurol. Scand.,* 2000, 102, 363-70.

Nguyen, DK; Botez, MI. Diaschisis and neurobehavior. *Can. J. Neurol. Sci.,* 1998, 25, 5-12.

Orr, HT; Chung, MY; Banfi, S; Kwiatkowski, TJ Jr; Servadio, A; Beaudet, AL; McCall, AE; Duvick, LA; Ranum, LP; Zoghbi, HY. Expansion of an unstable trinucleotide CAG repeat in spinocerebellar ataxia type 1. *Nature Genet.* 1993, 4, 221-6.

Pedraza, OL; Botez, MI. Thiamine status in inherited degenerative ataxias. *J. Neurol. Neursurg. Psychiatry,* 1992, 55, 136-7.

Ravizza, SM; McCormick, CA; Schlerf, JE; Justus, T; Ivry, RB; Fiez, JA. Cerebellar damage produces selective deficits in verbal working memory. *Brain,* 2006, 129, 306-20.

Rosenthal, G; Gilman, S; Koeppe, RA; Kluin, KJ; Markel, DS; Junck, L; Gebarski, SS. Motor dysfunction in olivopontocerebellar atrophy is related to cerebral metabolic rate studies with positron emission tomography. *Ann. Neurol.,* 1988, 24, 414-9.

Schmahmann, JD. An emerging concept: the cerebellar contribution to higher function. *Arch. Neurol.,* 1991, 48, 1178-87.

Schmahmann, JD; Sherman, JC. The cerebellar cognitive affective syndrome. *Brain,* 1998, 121, 561-79.

Seidel, K; Siswanto, S; Brunt, ER; den Dunnen, W; Korf, HW; Rüb, U. Brain pathology of spinocerebellar ataxias. *Acta Neuropathol.,* 2012, 124, 1-21.

Sonmezoglu, K; Sperling, B; Henriksen, T; Tfelt-Hansen, P; Lassen, NA. Reduced contralateral hemispheric flow measured by SPECT in cerebellar lesions: crossed cerebral diaschisis. *Acta Neurol. Scand.,* 1993, 87, 275-80.

Steinlin, M; Styger, M; Bolshauser, E. Cognitive impairments in patients with congenital nonprogressive cerebellar ataxia. *Neurology,* 1999, 53, 966-73.

Timmann, D; Daum, I. Cerebellar contributions to cognitive functions: a progress report after two decades of research. *Cerebellum,* 2007, 6, 159-62.

Timmann, D; Daum, I. How consistent are cognitive impairments in patients with cerebellar disorders? *Behav. Neurol.,* 2010, 23, 81-100.

Townsend, J; Courchesne, E; Covington, J; Westerfield, M; Harris, NS; Lyden, P; Lowry, TP; Press, GA. Spatial attention deficits in patients with acquired or developmental cerebellar abnormality. *J. Neurosci.,* 1999, 19, 5632-43.

Tsou, AY; Paulsen, EK; Lagedrost, SJ; Perlman, SL; Mathews, KD; Wilmot, GR; Ravina, B; Koeppen, AH; Lynch, DR. Mortality in Friedreich ataxia. *J. Neurol. Sci.,* 2011 307, 46-9.

White, M; Lalonde, R; Botez-Marquard, T. Neuropsychologic and neuropsychiatric characteristics of patients with Friedreich's ataxia. *Acta Neurol. Scand.,* 2000, 102, 222-6.

Zalesky, A; Akhlaghi, H; Corben, LA; Bradshaw, JL; Delatycki, MB; Storey, E; Georgiou-Karistianis, N; Egan, GF. Cerebello-cerebral connectivity deficits in Freidreich ataxia. *Brain Struct. Funct.,* 2014, 219, 969-81.

Yamaguchi, H; Fukuyama, H; Kimura, J. Hemodynamic and metabolic changes in crossed cerebellar hypoperfusion. *Stroke,* 1992, 23, 855-60.

In: Spatial Orientation in Humans and Animals
Editor: Robert Lalonde
ISBN: 978-1-63463-001-6
© 2014 Nova Science Publishers, Inc.

Chapter 9

SPATIAL DEFICITS IN ANIMAL MODELS OF CEREBELLAR DISORDERS

Robert Lalonde**[*1] **and Catherine Strazielle**[#2]
[1]Université de Rouen, Dépt. Psychologie,
Mont-Saint-Aignan Cedex France.
[2]Université de Lorraine, Laboratoire "Stress, Immunité, Pathogènes"
and Service de Microscopie Electronique, Faculté de Médecine and
CHU de Nancy, Vandoeuvre-les-Nancy, France

1. MORRIS WATER MAZE

The cerebellar contribution to cognitive processes in animals was first proposed by Watson (1978) and subsequently confirmed with respect to several aspects of behavior, including spatial orientation in different mazes performed by mutated mice and rats with surgical lesions (Lalonde, 1997).

1.1. Spontaneous Mutations

Several natural mutations with cerebellar cortical degeneration have been discovered (Table 1), which lead to ataxia (wide-spread gait) and dysmetria (imprecise movements). *Grid2*^{Lc} (*Lurcher*) and *Rora*^{sg} (*staggerer*) mutants are characterized by massive degeneration of cerebellar Purkinje and granule cells along with afferent regions such as the inferior olive without major atrophy elsewhere in the brain. The degeneration in *Grid2*^{ho} (*hot-foot*, also *Grid2*$^{ho\text{-}nancy}$ with the Nancy allele) is more discrete, affecting mostly granule cells. Likewise, *Agtpbp1*^{pcd} (*Purkinje cell degeneration*) and *nervous* are characterized by relatively selective damage to Purkinje cells, but also to retinal cells (Mullen and LaVail, 1975), though the latter

[*] Email: robert.lalonde@univ-rouen.fr
[#] Email: catherine.strazielle@univ-lorraine.fr

degeneration progresses more slowly. In the Dst^{dt-J} (*dystonia musculorum*) mutant, spinocerebellar tracts deteriorate, leading to crawling movements in addition to ataxia. These cerebellar mutants are affected in the Morris water maze (Table 2).

Table 1. Genetic and neuropathological features of behaviorally characterized cerebellar mutant mice

Mutant	Gene alteration	Cerebellar cell loss	References
$Grid2^{Lc}$	Autosomal semi-dominant *Grid2* malfunction	Purkinje cells, granule cells, deep nuclei	Caddy and Biscoe, 1979; Zuo et al., 1997
$Grid2^{ho}$	Autosomal recessive *Grid2* deletion	granule cells	Guastavino et al., 1990; Lalouette et al., 1998
$Rora^{sg}$	Autosomal recessive *Rora* deletion	Purkinje cells granule cells	Hamilton et al., 1996; Herrup, 1983; Herrup and Mullen, 1979; Landis and Sidman, 1978; Roffler-Tarlov and Herrup, 1981; Sotelo, 1975
$Agtpbp1^{pcd}$	Autosomal recessive *Agtpbp1* deletion	Purkinje cells, deep nuclei	Fernandez-Gonzalez et al., 2002; Ghetti et al., 1987; Mullen et al., 1976; Triarhou and Ghetti, 1991; Triarhou et al., 1987
nervous	Autosomal recessive unkown gene	Purkinje cells	Sidman and Green, 1970; Sotelo and Triller, 1979
Dst^{dt-J}	Autosomal recessive *Dst* deletion	Spinocerebellar tracts	Brown et al., 1995; Sotelo and Guénet, 1988

Table 2. The acquisition of hidden and visible platform versions of the Morris water maze by cerebellar mutant mice

Mutant	Hidden platform	Visible platform	References
$Grid2^{Lc}$	↓	↓	Cendelín et al., 2008; Hilber et al., 1998; Lalonde and Thifault, 1994; Lalonde et al., 1988a
$Rora^{sg}$	↓	normal	Lalonde, 1987
$Agtpbp1^{pcd}$	↓	normal	Goodlett et al., 1991
nervous	↓	normal	Lalonde and Strazielle, 2003
Dst^{dt-J}	?	↓	Lalonde et al., 1993

↓=deficit, normal=no deficit, ?=not yet tested.

The loss in balance and equilibrium noticeable in cerebellar mutants moving on dry ground is less apparent in a buoyant medium. The animals swim efficiently and without sinking provided trial durations are not excessively prolonged. Despite massive atrophy of the cerebellar cortex, rhythmic movements used for swimming were minimally affected in *Grid2^{Lc}* mutants relative to non-mutated mice of the same background strain (Fortier et al., 1987) and their swim speed was the same (Cendelín et al., 2008). Such a profile facilitates interpretations as to the cognitive impact of cerebellar damage independently of motor control.

During acquisition of the hidden platform subtask of the Morris water maze (chapter 2), distance swum and escape latencies increased in *Rora^{sg}* (Lalonde, 1987), *Agtpbp1^{pcd}* (Goodlett et al., 1991), and *nervous* (Lalonde and Strazielle, 2003) mutants compared with age-matched controls, whereas the same measures were not affected in the visible platform subtask. These results indicate a specific deficit in spatial orientation as opposed to the visuomotor guidance required to swim towards a cued target. *Agtpbp1^{pcd}* and *nervous* mutants were evaluated at an age with miminal or nonexistent retinal degeneration, thus minimizing the possibility of a loss in visual acuity as the underlying cause of the disssociation. Moreover, the similar profile exhibited by the *Rora^{sg}* mutant, never shown to be visually impaired, is concordant with the hypothesis that cerebellar damage can cause a selective loss in spatial orientation.

The deficits revealed by the *Grid2^{Lc}* mutant were more severe, encompassing both submerged and visible platform tests (Cendelín et al., 2008; Hilber et al., 1998; Lalonde and Thifault, 1994; Lalonde et al., 1988a). Thus, cerebellar damage may sometimes cause a disturbance in visuomotor guidance despite near normal swimming movements (Fortier et al., 1987). The more severe phenotype seen in *Grid2^{Lc}* relative to *Agtpbp1^{pcd}* and *nervous* may be due to more widespread cerebellar atrophy. The more severe phenotype in *Grid2^{Lc}* relative to *Rora^{sg}* may be due to the earlier onset of cerebellar atrophy in the latter, causing increased susceptibility to compensatory responses. Although oculomotor deficits have been described in *Grid2^{Lc}* (Stahl et al., 2002) akin to the oculomotor dysmetria displayed in patients with cerebellar atrophy (Filippopulos et al., 2013; Moschner et al., 1999; Thier et al., 1998), it is doubtful that this sign has an impact on maze behavior, because rodents generally explore the environment by moving their heads rather than their eyes.

A different phenotype was discerned in the *Dst$^{dt\text{-}J}$* spinocerebellar mutant (Lalonde et al., 1993). Indeed, *Dst$^{dt\text{-}J}$* mutants are more liable to sink than cerebellar mutants to the extent of precluding evaluation in the hidden platform test. *Dst$^{dt\text{-}J}$* mutants had longer escape latencies than controls but shorter distance swum before reaching the visible platform, indicating slower swim speed. Shorter distances may be explained by difficulties in maintaining their bodies above water level, thereby reducing the temptation of exploring the maze as widely as controls. These results imply that the loss in directional control seen in *Grid2^{Lc}* mutants is probably mediated not from the spinal cord but by afferents originating from inferior olive, reticular or pontine nuclei.

1.2. Induced Mutations

Transgenic L7PKCI mice are characterized by disruption of protein kinase C (PKC)-dependent long-term depression (LTD), an electrophysiological index of plasticity, at parallel fiber-Purkinje cell synapses (De Zeeuw et al., 1998; Oberdick et al., 1990). L7PKCI mutants acquired the hidden platform task of the Morris water maze more slowly than non-transgenic controls (Burguière et al., 2005). In contrast, there was no intergroup difference on the visible platform task. L7PKCI mutants were also impaired relative to non-transgenic controls in the path integration task (chapter 2) (Rochefort et al., 2011). The properties of hippocampal place cells were solely impaired when L7PKCI mice relied on self-motion cues. These results show that cerebellar PKC-dependent mechanisms are involved in processing self-motion signals essential to spatial representations in the hippocampus (Rochefort et al., 2013; Rondi-Reig and Burguière, 2005). A neurocomputational model using Burguière et al's (2005) experimental data supported the notion of cerebellar involvement in spatial information (Passot et al., 2012).

1.3. Surgical Lesions

Surgical lesioning of the cerebellum provides complementary information to the analysis of mutant behavior, in particular regarding the selectivity of brain regions responsible for spatial orientation. Bilateral aspiration lesions of the cerebellar cortex (Goddyn et al., 2006) or excitotoxic lesions limited to the right cerebellar hemisphere (Edalatmanesh et al., 2011) slowed down acquisition of the hidden platform task. The same deficit occurred after bilateral aspiration lesions limited to the lateral cerebellar cortex and associated nuclei or fastigial nucleus alone (Joyal et al., 1996), electrolytic lesions of the dentate nucleus (Joyal et al., 2001; Noblett and Swain, 2003), or intracerebroventricular injections of OX-7-saporin, a neurotoxin that selectively destroyed 56% of Purkinje cells (Gandhi et al., 2000). In contrast, colchicine-induced lesions of dentate nuclei had no effect on acquisition of an egocentric version where the hidden platform was always situated on the same side as where rats began the trial (Gaytán-Tocavén and Olvera-Cortés, 2004), indicating that the previous impairments were caused by deficient processing of allocentric cues. In an adaptation of the task with rats placed from the same starting point instead of different ones, acquisition was slowed down after cerebellectomy (Hilber et al., 1998) or in a granulo-deprived cerebellum caused by X-irradiation prior to weaning and tested at 3 months of age (Le Marec et al., 1997).

Unlike their deleterious effects on the hidden platform task, lesions of lateral cerebellar cortex and associated nuclei (Joyal et al., 1996), dentate nucleus (Joyal et al., 2001), fastigial nucleus (Joyal et al., 1996), or Purkinje cells (Gandhi et al., 2000) had no effect on the visible platform task. These data provide evidence that spatial orientation is dependent on the integrity of cerebellar cortex and nuclei.

In contrast to lesions of lateral cerebellar cortex or deep cerebellar nuclei, lesions of vermal cortex and fastigial nucleus impaired both submerged and visible platform tasks (Joyal et al., 1996). Likewise, visible platform performance was impaired after unilateral cerebellectomy (Colombel et al., 2004; Petrosini et al., 1996). Unlike sham-operated controls, cerebellectomized rats displayed persistent thigmotaxic behavior (exaggerated turning around the pool perimeter), though not necessarily on the ipsilateral side of the ablation (Petrosini et al., 1996). On the contrary, performance was not deficient when the task was learned prior to the lesion. These results were interpreted as an impairment in procedural learning, as defined by disinhibition of the prepotent wall-hugging response seen in normal rats during early stages of training. Indeed, hemicerebellectomized rats also displayed a lack in behavioral flexibility during a two-step hidden platform task with black curtains surrounding the water maze and a shifting platform location which were then removed (Leggio et al., 1999). In addition, hemicerebellectomized rats were impaired in observational learning of the Morris water maze (Leggio et al., 2000). As long as the hemicerebellectomized rats were exposed to the experimental situation either through prior training (Petrosini et al., 1996) or observation of demonstrator rats (Graziano et al. 2002; Leggio et al., 2000), no deficit was discerned, giving credence to the hypothesis that poor procedural learning constitutes one factor underlying the visuomotor deficit (Molinari and Leggio, 2007; Petrosini et al., 1998).

Overall, these results indicate that lateral cerebellar cortex and deep cerebellar nuclei contribute to the neural circuitry underlying spatial orientation in conjunction with the telencephalon. In contrast, midline cerebellar cortex and fastigial nucleus are essential for visuomotor guidance.

1.4. Surgical Lesions or Pharmacologic Manipulation of Cerebellar Afferents

Either 3-acetylpyridine-induced lesions of the inferior olive, destroying climbing fibers, or electrolytic lesions of the medial cerebellar peduncle, destroying mossy fibers, slowed down acquisition of the hidden platform task in adult rats (Gasbarri et al., 2003). The same deficit was found in the adapted version from the same starting point after 3-acetylpyridine-induced lesions of the inferior olive (Dahhaoui et al., 1992), or after 3-acetylpyridine combined with X-irradiation administered during the pre-weaned period in rats tested as adults (Rondi-Reig et al., 2002). The latter procedure had no effect on visible platform performance. Delayed acquisition of the hidden platfrom task was also observed after lidocaine injections into the locus coeruleus (Khakpour-Taleghani et al., 2009), possibly due in part to a disrupted coeruleo-cerebellar pathway. Moreover, harmaline-induced stimulation of the inferior olive impaired both hidden and visible platform versions (Meignin et al., 1999). In contrast, unilateral transection of the inferior cerebellar peduncle on postnatal day 3 or 11 impaired the hidden but not the visible platform task in juvenile rats (Willson et al., 2007). Thus, as with cerebellar lesions, lesions of afferent structures cause deficits in either or both platform tasks depending on lesion type.

2. Multi-Unit Mazes

2.1. Mutant Mice

The spatial deficits seen in cerebellar mutant mice swimming in the Morris water maze are prevalent as well in other mazes. *Grid2*^{Lc}, *Grid2*$^{ho\text{-}nancy}$, and *Rora*^{sg} mutant mice had deficient route-finding abilities in the Z-maze (chapter 2) relative to non-ataxic controls (Filali et al., 1996; Lalonde et al., 1996).

2.2. Surgical Lesions and X-Irradiation

X-irradiation of the cerebellum was applied during pre-weaning stages in rats evaluated in a multi-unit maze at 60 days of age (Pellegrino and Altman, 1979). When X-irradiation was applied during postnatal days 12 to 15, causing agenesis of late-forming granule cells, or during postnatal days 4 to 5, causing morphological disorganization of the cerebellar cortex with intermediate cell agenesis, or during postnatal days 4 to 15, causing a disorganized cerebellar cortex nearly devoid of all postnatally forming interneurons, the rats were slowed down during acquisition of the multi-unit water maze. The authors concluded that the role of the cerebellum is not limited to motor coordination but extends to the coordination of mental processes. In confirmation of this hypothesis is the finding that performance of a preoperatively acquired 3-unit maze was impaired after large cerebellar lesions (Thompson, 1974).

3. Left-Right Discrimination

Hemicerebellectomized rats were slower than sham controls in acquiring a left-right discrimination in a water-filled T-maze (Molinari et al., 1997; Petrosini et al., 1996). In contrast, *Grid2*^{Lc} mutant mice were not impaired in the same task (Belzung et al., 2001; Lalonde et al., 1988a).

Moreover, bilateral cerebellar cortical ablation did not affect spatial reversal learning for food reward in rats (Thompson, 1983). It appears that only complete destruction of the cerebellum, though only on one side, is necessary to impair spatial discrimination and perhaps its reversal.

4. Radial-Arm Maze

In contrast to the above-named reference memory tasks, the radial-arm maze is used to estimate working memory (chapter 2). Food-deprived *Rora*^{sg} (Goldowitz and Koch, 1986) and *Grid2*^{Lc} (Belzung et al., 2001) mutants committed more errors than controls in the radial-arm maze task. Hemicerebellectomized rats were also slower in acquiring the radial-arm maze task, but were not affected when performance of the already acquired task was examined (Mandolesi et al., 2001). More specific colchicine-induced lesions of the dentate nucleus slowed down the acquisition of the radial-arm maze task (Gaytán-Tocavén and Olvera-Cortés, 2004). Since electrical stimulation of the vestibular system impaired radial maze performance, one may postulate a role for vestibulo-cerebellar interactions underlying this task (Aversano et al., 2002).

5. Reinforced Alternation

Rora^{sg} mutant mice committed more errors while acquiring left-right alternations in a water maze (Lalonde, 1987). But unlike aspiration lesions of prefrontal cortex, spatial alternation performance was not affected after excitotoxic lesions of the dentate nucleus in monkeys (Nixon and Passingham, 1999).

6. Spontaneous Alternation

Spontanous alternation deficits have been noted in several cerebellar mutants, including *nervous* (Lalonde et al., 1986a), *Agtpbp1*^{pcd} (Lalonde et al., 1987), *Grid2*^{Lc} (Caston et al., 1997; Lalonde et al., 1986b), *Grid2*$^{ho\text{-}nancy}$ (Filali et al., 1996), and *Rora*^{sg} (*staggerer*) (Caston et al., 2004; Lalonde et al., 1988b). On the contrary, no such deficit was reported in the *Dst*$^{dt\text{-}J}$ spinocerebellar mutant (Lalonde et al., 1994), a result that underlines a specific role to the cerebellum on spatial or emotion-related processes as opposed to the motor handicap. Spontanous alternation deficits have also been reported in cerebellectomized mice (Caston et al., 1997) as well as cerebellar X-irradiated rats (Pellegrino and Altman, 1979).

7. Cerebellar-Related Pathways Underlying Spatial Orientation

Stein (1986) first proposed a role for the cerebellum in the visual guidance of movement. This hypothesis concords with findings of deficient spatial orientation and visuomotor guidance in the Morris water maze as well as other maze tasks in mice or rats with cerebellar lesions.

Because vestibular input is involved in the radial-arm maze (Aversano et al., 2002) and the Morris water maze (Semenov and Bures, 1989), one may infer that bidirectional vestibulo-cerebellar interactions underlie the directional control of body movements through space. Vestibular cues in the Morris task seem particularly vital in orienting the animal in the gravitation field and estimating angles between vectors plotted from the body towards external landmarks (Semenov and Bures, 1989). The concordance between vestibular and visual signals appears to be a prerequisite for efficient navigation.

In addition to vestibulo-cerebellar interactions, the hippocampus has often been demonstrated to be crucial for place learning though not for visuomotor guidance in the Morris task (chapter 2). The likelihood that the hippocampus uses vestibulo-cerebellar signals for efficient navigation is supported by findings that hippocampal place cells are dependent on vestibular (Stackman et al., 2002) and cerebellar (Rochefort et al., 2011) input. Vestibular nuclei are not known to reach the hippocampus directly, but may do so via entorhinal and perirhinal cortex (Rubertone et al., 1995). The cerebellum may also mediate vestibular input to the medial temporal lobe via indirect projections to the hippocampus (Heath et al., 1978; Newman and Reza, 1979). In return, the hippocampus sends information back to the cerebellar cortex via brainstem nuclei projecting climbing and mossy fibers (Saint-Cyr and Woodward, 1980).

Another manner by which the cerebellum may modulate spatial orientation is via the cerebello-thalamo-cortical pathway (chapter 3). The cerebellum projects through ventrolateral thalamic nuclei not only to motor cortex but also to parietal (Amino et al., 2001; Gianetti and Molinari, 2002) and frontal (Middleton and Strick, 2001) association cortices, which in turn send information back to the cerebellar cortex via basal pontine nuclei (Schmahmann and Pandya, 1997). In support of the involvement for cerebello-cortical interactions in spatial orientation is the finding that lesions of the ventrolateral/ventromedial nuclei of the thalamus, the main cerebellar efferent relay station to the cerebral cortex, cause hidden platform deficits in the Morris water maze (Jeljeli et al., 2003), as do lesions of posterior parietal and medial prefrontal cortex (chapter 2).

The cerebellum may also affect spatial orientation by communicating with the basal ganglia. Output of the dentate nucleus reaches the striatum via a disynaptic pathway in the monkey (Hoshi et al., 2005), possibly via midbrain dopaminergic neurons, targets of cerebellar projections in the rat (Perciavalle et al., 1989; Snider, 1976). In addition, it has been proposed that the vestibular system contributes to spatial orientation via the neostriatum (Potegal, 1982).

The effects of cerebellar lesions in animals support findings of visuospatial deficits found in patients with cerebellar atrophy (chapter 8). In a wider view, the role of the cerebellum on spatial functions may be integrated in the context of reinforcement learning (Swain et al., 2011).

References

Amino, Y; Kyuhou, S; Matzuzaki, R; Gemba, H. Cerebello-thalamo-cortical projections to the posterior parietal cortex in the macaque monkey. *Neurosci. Lett.,* 2001, 309, 29-32.

Aversano, M; Sacchi, R; Memoli, R; Graziano, A; Petrosini, L. Anterograde and retrograde influence of vestibular stimulation on spatial working memory. *Physiol. Behav.,* 2002, 76, 655-60.

Belzung, C; Chapillon, P; Lalonde, R. The effects of the *Lurcher* mutation on object localization, T-maze left-right discrimination, and radial arm maze tasks. *Behav. Genet.,* 2001, 31, 151-5.

Brown, A; Berthier, G; Mathieu, M; Rossant, J; Kothary, R. The mouse *dystonia musculorum* gene is a neural isoform of *bullous pemphigoid antigen 1. Nature Genet.,* 1995, 10, 301-6.

Burguière, E; Arleo, A; Hojjati, Mr; Elgersma, Y; De Zeeuw, CI; Berthoz, A; Rondi-Reig, L. Spatial navigation impairment in mice lacking cerebellar LTD: a motor adaptation deficit? *Nature Neurosci.,* 2005, 8, 1292-4.

Caddy, KWT; Biscoe, TJ. Structural and quantitative studies on the normal C3H and *Lurcher* mutant mouse. *Philos. Trans. Roy. Soc. Lond. (Biol.),* 1979, 287, 167-201.

Caston, J; Chianale, C; Mariani, J. Spatial memory of heterozygous staggerer (Rora(+)/Rora(*sg*) versus normal (Rora(+)/Rora(+) mice during aging. *Behav. Genet.,* 2004, 34, 319-24.

Caston, J; Vasseur, F; Delhaye-Bouchaud, N; Mariani, J. Delayed spontaneous alternation in intact and cerebellectomized control and lurcher mutant mice: differential role of cerebellar cortex and deep cerebellar nuclei. *Behav. Neurosci.,* 1997, 111, 214-8.

Cendelín, J; Korelusová, I; Vozeh, F. The effect of repeated rotarod training on motor skills and spatial learning ability in Lurcher mutant mice. *Behav. Brain Res.,* 2008, 189, 65-74.

Colombel, C; Lalonde, R; Caston, J. The effects of unilateral removal of the cerebellar hemispheres on spatial learning and memory in rats. *Brain Res.,* 2004, 1004, 108-15.

Dahhaoui, M; Stelz, T; Caston, J. Effects of lesion of the inferior olivary complex by 3-acetylpyridine on learning and memory in the rat. *J. Comp. Physiol. A,* 1992, 171, 657-64.

De Zeeuw, CI; Hansel, C; Bian, F; Koekkoek, SK; van Alphen, AM; Linden, DJ; Oberdick, J. Expression of a protein kinase C inhibitor in Purkinje cells blocks cerebellar LTD and adaptation of the vestibulo-ocular reflex. *Neuron,* 1998, 20, 495-508.

Edalatmanesh, MA; Bahrami, AR; Hosseini, E; Hosseini, M; Khatamsaz, S. Bone marrow derived mesenchymal stem cell transplantation in cerebellar degeneration: a behavioral study. *Behav. Brain Res.,* 2011, 225, 63-70.

Fernandez-Gonzalez, A; La Spada, A; Treadaway, J; Higdon, JC; Harris, BS; Sidman, RL; Morgan, JI; Zuo, J. *Purkinje cell degeneration* (*pcd*) phenotypes caused by mutations in the axotomy-induced gene, *Nna1*. *Science,* 2002, 295, 1904-8.

Filali, M; Lalonde, R; Bensoula, AN; Guastavino, J-M; Lestienne, F. Spontaneous alternation, motor activity, and spatial learning in *hot-foot* mutant mice. *J. Comp. Physiol. A,* 1996, 178, 101-4.

Filippopulos, F; Eggert, T; Straube, A. Deficits of cortical oculomotor mechanisms in cerebellar atrophy patients. *Exp. Brain Res.,* 2013, 224, 541-50.

Fortier, P; Rossignol, S; Smith, AM. Locomotor deficits in the cerebellar mutant mouse, *Lurcher*. *Exp. Brain Res.,* 1987, 66, 271-86.

Gandhi, CC; Kelly, RM; Wiley, RG; Wash, TJ. Impaired acquisition of a Morris water maze task following selective destruction of cerebellar Purkinje cells with OX7-saporin. *Behav. Brain Res.,* 2000, 109, 37-47.

Gasbarri, A; Pompili, A; Pacitti, C; Cicirata, F. Comparative effects of lesions to the ponto-cerebellar and olivo-cerebellar pathways on motor and spatial learning in the rat. *Neuroscience,* 2003, 116, 1131-40.

Gaytán-Tocavén, L; Olvera-Cortés, ME. Bilateral lesion of the cerebellar-dentate nucleus impairs egocentric sequential learning but not egocentric navigation in the rat. *Neurobiol. Learn. Mem.,* 2004, 82, 120-7.

Ghetti, B; Norton, J; Triarhou, LC. Nerve cell atrophy and loss in the inferior olivary complex of "*Purkinje cell degeneration*" mutant mice. *J. Comp. Neurol.,* 1987, 260, 409-22.

Gianetti, S; Molinari, M. Cerebellar input to the posterior parietal cortex in the rat. *Brain Res. Bull.,* 2002, 58, 481-9.

Goddyn, H; Leo, S; Meert, T; D'Hooge, R. Differences in behavioural test battery performance between mice with hippocampal and cerebellar lesions. *Behav. Brain Res.,* 2006, 173, 138-47.

Goldowitz, D; Koch, J. Performance of normal and neurological mutant mice on radial maze and active avoidance tests. *Behav. Neur. Biol.,* 1986, 46, 216-26.

Goodlett, RC; Hamre, KM; West, JR. Dissociation of spatial navigation and visual guidance performance in *Purkinje cell degeneration* (*pcd*) mutant mice. *Behav. Brain Res.,* 1991, 47, 129-41.

Graziano, A; Leggio, MG; Mandolesi, L; Neri, P; Molinari, M; Petrosini, L. Learning power of single behavioral units in acquisition of a complex spatial behavior: an observational learning study in cerebellar-lesioned rats. *Behav. Neurosci.,* 2002, 116, 116-25.

Guastavino, J-M; Sotelo, C; Damez-Kinselle, I. *Hot-foot* murine mutation: behavioral effects and neuroanatomical alterations. *Brain Res.,* 1990, 523, 199-210.

Hamilton, BA; Frankel, WN; Kerrebrock, AW; Hawkins, TL; FitzHugh, W; Kusumi, K; Russell, LB; Mueller, KL; van Berkel, V; Birren, BW; Kruglyak, L; Lander, ES. Disruption of the nuclear hormone receptor RORalpha in *staggerer* mice. *Nature,* 1996, 379, 736-9.

Heath, RG; Dempesy, CW; Fontana, CJ; Myers, WA. Cerebellar stimulation: effects on septal region, hippocampus, and amygdala of cats and rats. *Biol. Psychiatry,* 1978, 13, 501-29.

Herrup, K. Role of *staggerer* gene in determining cell number in cerebellar cortex. I. Granule cell death is an indirect consequence of *staggerer* gene action, *Dev. Brain Res.,* 1983, 11, 267-74.

Herrup, K; Mullen, RJ. Regional variation and absence of large neurons in the cerebellum of the *staggerer* mouse. *Brain Res.,* 1979, 172, 1-12.

Hilber, P; Jouen, F; Delhaye-Bouchaud, N; Mariani, J; Caston, J. Differential roles of cerebellar cortex and deep cerebellar nuclei in learning and retention of a spatial task: studies in intact and cerebellectomized *Lurcher* mutant mice. *Behav. Genet.,* 1998, 28, 299-308.

Hoshi, E; Tremblay, L; Féger, J; Carras, PL; Strick, PL. The cerebellum communicates with the basal ganglia. *Nature Neurosci.,* 2005, 8, 1491-3.

Jeljeli, M; Strazielle, C; Caston J; Lalonde, R. Effects of ventrolateral-ventromedial thalamic lesions on motor coordination and spatial orientation in rats. *Neurosci. Res.,* 2003, 47, 309-16.

Joyal, CC; Strazielle, C; Lalonde, R. Effects of dentate nucleus lesions on spatial and postural sensorimotor learning in rats. *Behav. Brain Res.,* 2001, 122, 131-7.

Joyal, CC; Meyer, C; Jacquart, G; Mahler, P; Caston, J; Lalonde, R. Effects of midline and lateral cerebellar lesions on motor coordination and spatial orientation. *Brain Res.,* 1996, 739, 1-11.

Khakpour-Taleghani, B; Lashgari, R; Motamedi, F; Naghdi, N. Effect of reversible inactivation of locus ceruleus on spatial reference and working memory. *Neuroscience,* 2009, 158, 1284-91.

Lalonde, R. Exploration and spatial learning in *staggerer* mutant mice. *J. Neurogenet.,* 1987, 4, 285-92.

Lalonde, R. Visuospatial functions. In: JD Schmahmann (ed) *The cerebellum and cognition* (International Review of Neurobiology, vol 41). San Diego: Academic Press, pp 191-215, 1997.

Lalonde, R; Strazielle, C. Motor coordination, exploration, and spatial learning in a natural mouse mutation (*nervous*) with Purkinje cell degeneration. *Behav. Genet.,* 2003, 33, 59-66.

Lalonde, R; Thifault, S. Absence of an association between motor coordination and spatial orientation in *Lurcher* mutant mice. *Behav. Genet.,* 1994, 24, 497-501.

Lalonde, R; Botez, MI; Boivin, D. Spontaneous alternation and habituation in a t-maze in *nervous* mutant mice. *Behav. Neurosci.,* 1986a, 100, 350-2.

Lalonde, R; Joyal, CC; Botez, MI. Exploration and motor coordination in *dystonia musculorum* mutant mice. *Physiol. Behav.,* 1994, 56, 277-80.

Lalonde, R; Joyal, CC; Côté, C. Swimming activity in *dystonia musculorum* mutant mice. *Physiol. Behav.,* 1993, 54, 119-20.

Lalonde, R; Lamarre, Y; Smith, AM. Does the mutant mouse *Lurcher* have deficits in spatially oriented behaviours? *Brain Res.,* 1988a, 455, 24-30.

Lalonde, R; Manseau, M; Botez, MI. Spontaneous alternation and exploration in *staggerer* mutant mice. *Behav. Brain Res.,* 1988b, 27, 273-6.

Lalonde, R; Manseau, M; Botez, MI. Spontaneous alternation and habituation in *Purkinje cell degeneration* mutant mice. *Brain Res.,* 1987, 411, 187-9.

Lalonde, R; Lamarre, Y; Smith, AM; Botez, MI. Spontaneous alternation and habituation in *Lurcher* mutant mice. *Brain Res.,* 1986b, 362, 161-4.

Lalonde, R; Filali, M; Bensoula, AN; Monnier, C; Guastavino, J-M. Spatial learning in a Z-maze by cerebellar mutant mice. *Physiol. Behav.,* 1996, 59, 83-6.

Lalouette, A; Guénet, J-L; Vriz, S. *Hot-foot* mutations affect the δ2 glutamate receptor gene and are allelic to *Lurcher. Genomics,* 1998, 50, 9-13.

Landis, DMD; Sidman, RL. Electron microscopic analysis of postnatal histogenesis in the cerebellar cortex of *staggerer* mutant mice. *J. Comp. Neurol.,* 1978, 179, 831-64.

Leggio, MG; Molinari, M; Neri, P; Graziano, A; Mandolesi, L; Petrosini, L. Representation of actions in rats: the role of cerebellum in learning spatial performances by observation. *Proc. Natl. Acad. Sci. USA,* 2000, 97, 2320-5.

Leggio, MG; Neri, P; Graziano, A; Mandolesi, L; Molinari, M; Petrosini, L. Cerebellar contribution to spatial event processing: characterization of procedural learning. *Exp. Brain Res.*, 1999, 127, 1-11.

Le Marec, N; Dahhaoui, M; Stelz, T; Bakalian, A; Delhaye-Bouchaud, N; Caston, J; Mariani, J. Effect of cerebellar granule cell depletion on spatial learning and memory and in an avoidance conditioning task: studies in potnatally X-irradiated rats. *Dev. Brain Res.,* 1997, 99, 20-8.

Mandolesi, L; Leggio, MG; Graziano, A; Neri, P; Molinari, M. Cerebellar contribution to spatial event processing: involvement in procedural and working memory components. *Eur. J. Neurosci.,* 2001, 14, 2011-22.

Meignin, C; Hilber, P; Caston, J. Influence of stimulation of the olivocerebellar pathway by harmaline on spatial learning in the rat. *Brain Res.,* 1999, 824, 277-83.

Middleton, FA; Strick, PL. Cerebellar projections to the prefrontal cortex of the primate. *J. Neurosci.,* 2001, 21, 700-12.

Molinari, M; Grammaldo, LG; Petrosini, L. Cerebellar contribution to spatial event processing: right-left discrimination abilities in rats. *Eur. J. Neurosci.,* 1997, 9, 1986-97.

Molinari, M; Leggio, MG. Cerebellar information processing and visuospatial functions. *Cerebellum,* 2007, 6, 214-20.

Moschner, C; Crawford, TJ; Heide, W; Trillenberg, P; Kömpf, D; Kennard, C. Deficits of smooth pursuit initiation in patients with degenerative cerebellar lesions. *Brain,* 1999, 122, 2147-58.

Mullen, RJ; LaVail, MM. Two new types of retinal degeneration in cerebellar mutant mice. *Nature,* 1975, 258, 528-30.

Mullen, RJ; Eicher, EM; Sidman, RL. *Purkinje cell degeneration*: a new neurological mutation in the mouse. *Proc. Natl. Acad. Sci. USA,* 1976, 73, 208-12.

Newman, PP; Reza, H. Functional relationships between the hippocampus and the cerebellum: an electrophysiological study of the cat. *J. Physiol.,* 1979, 287, 405-26.

Nixon, PD; Passingham, RE. The cerebellum and cognition: cerebellar lesions do not impair spatial working memory or visual associative learning in monkeys. *Eur. J. Neurosci.,* 1999, 11, 4070-80.

Noblett, KL; Swain, RA. Pretraining enhances recovery from visuospatial deficit following cerebellar dentate nucleus lesion. *Behav. Neurosci.,* 2003, 117, 785-98.

Oberdick, J; Smeyne, RJ; Mann, JR; Zackson, S; Morgan, JI. A promoter that drives transgene expression in cerebellar Purkinje and retinal bipolar neurons. *Science,* 1990, 248, 223-6.

Passot, JB; Sheynikhovich, D; Duvelle, É; Arleo, A. Contribution of cerebellar sensorimotor adaptation to hippocampal spatial memory. *PLoS One,* 2012, 7(4), e32560.

Pellegrino, LJ; Altman, J. Effects of differential interference with postnatal cerebellar neurogenesis on motor performance, activity level, and maze learning of rats: a developmental study. *J. Comp. Physiol. Psychol.,* 1979, 93, 1-33.

Perciavalle, V; Berretta, S; Raffaele, R. Projections from the intracerebellar nuclei to the ventral midbrain tegmentum in the rat. *Neuroscience,* 1989, 29, 109-19.

Petrosini, L; Leggio, MG; Molinari, M. The cerebellum in the spatial problem solving: a co-star or a guest star? *Prog. Neurobiol.,* 1998, 56, 191-210.

Petrosini, L; Molinari, M; Dell'Anna, ME. Cerebellar contribution to spatial event processing: Morris water maze and T-maze. *Eur. J. Neurosci.,* 1996, 8, 1882-96.

Potegal, M. Vestibular and neostriatal contributions to spatial orientation. In: M Potegal (ed) *Spatial abilities: Development and physiological foundations.* New York: Academic Press, pp 361-87, 1982.

Rochefort, C; Lefort, JM; Rondi-Reig, L. The cerebellum: a key structure in the navigation system. *Front. Neur. Circuits,* 2013, 7, 35.

Rochefort, C; Arabo, A; André, M; Poucet, B; Save E; Rondi-Reig, L. Cerebellum shapes hippocampal spatial code. *Science,* 2011, 334, 385-9.

Roffler-Tarlov, S; Herrup, K. Quantitative examination of the deep cerebellar nuclei in the *staggerer* mutant mouse. *Brain Res.,* 1981, 215, 49-59.

Rondi-Reig, L; Burguière, E. Is the cerebellum ready for navigation? *Prog. Brain Res.,* 2005, 148, 199-212.

Rondi-Reig, L; Le Marec, N; Caston, J; Mariani, J. The role of climbing and parallel fibers inputs to cerebellar cortex in navigation. *Behav. Brain Res.,* 2002, 132, 11-8.

Rubertone, JA; Mehler, WR; Voogd, J. The vestibular nuclear complex. In: G Paxinos (ed) *The rat nervous system* (2nd edition). New York: Academic Press, pp 773-796, 1995.

Saint-Cyr, JA; Woodward, DJ. Activation of mossy and climbing fiber pathways to the cerebellar cortex by stimulation of the fornix in the rat. *Exp. Brain Res.,* 1980, 40, 1-12.

Schmahmann, JD; Pandya, DN. Anatomic organization of the basilar pontine projections from prefrontal cortices in rhesus monkey. *J. Neurosci.,* 1997, 17, 438-58.

Semenov, LV; Bures, J. Vestibular stimulation disrupts acquisition of place navigation in the Morris water tank task. *Behav. Neural Biol.,* 1989, 51, 346-63.

Sidman, RL; Green, MC. "*Nervous*", a new mutant mouse with cerebellar disease. In: M Sabourdy (ed) *Les mutants pathologiques chez l'animal.* Paris: CNRS, pp 69-79, 1970.

Snider, RS; Maiti, A; Snider, SR. Cerebellar pathways to ventral midbrain and nigra. *Exp. Neurol.,* 1976, 53, 714-28.

Sotelo, C. Dendritic abnormalities of Purkinje cells in the cerebellum of neurologic mutant mice (*weaver* and *staggerer*). In: GW Kreutzberg (ed) *Physiology and pathology of dendrite.* New York: Raven Press, pp 335-51, 1975.

Sotelo, C; Guénet, J-L. Pathologic changes in the CNS of *dystonia musculorum* mutant mouse: an animal model for human spinocerebellar ataxia. *Neuroscience,* 1988, 27, 403-24.

Sotelo, C; Triller, A. Fate of presynaptic afferents of Purkinje cells in adult *nervous* mutant mice: a model to study presynaptic stabilization. *Brain Res.,* 1979, 175, 11-36.

Stahl, JS. Calcium channelopathy mutants and their role in oculomotor research. *Ann. NY Acad. Sci.,* 2002, 956, 64-74.

Stackman, RW; Clark, AS; Taube, JS. Hippocampal representations require vestibular input. *Hippocampus,* 2002, 12, 291-303.

Stein, JF. Role of the cerebellum in the visual guidance of movement. *Nature,* 1986, 323, 217-21.

Swain, RA; Kerr, AL; Thompson, RF. The cerebellum: a neural system for the study of reinforcement learning. *Front. Behav. Neurosci.,* 2011, 5, 8.

Thier, P; Haarmeier, T; Treue, S; Barash, S. Absence of a common functional denominator of visual disturbances in cerebellar disease. *Brain,* 1998, 122, 2133-46.

Thompson, R. Localization of the "maze memory system" in the white rat. *Physiol. Psychol.,* 1974, 2, 1-17.

Thompson, R. Abnormal learning and forgetting of individual spatial reversal problems in brain-damaged rats. *Physiol. Psychol.,* 1983, 11, 35-46.

Triarhou, LC. Rate of neuronal fallout in a transsynaptic cerebellar model. *Brain Res. Bull.,* 1998, 47, 219-22.

Triarhou, LC; Ghetti, B. Stabilisation of neurone number in the inferior olivary complex of aged "*Purkinje cell degeneration*" mutant mice. *Acta Neuropathol.,* 1991, 81, 597-602.

Triarhou, LC; Norton, J; Ghetti, B. Anterograde transsynaptic degeneration in the deep cerebellar nuclei of *Purkinje cell degeneration* mutant mice. *Exp. Brain Res.,* 1987, 66, 577-88.

Watson, PJ. Nonmotor functions of the cerebellum. *Psychol. Bull.,* 1978, 85, 944-67.

Willson, ML; Bower, AJ; Sherrard, RM. Developmental neural plasticity and its cognitive benefits: olivocerebellar reinnervation compensates for spatial function in the cerebellum. *Eur. J. Neurosci.,* 2007, 25, 1475-83.

Zuo, J; De Jager, PL; Takahashi, KA; Jiang, W; Linden, DJ; Heintz, N. Neurodegeneration in *Lurcher* mutant mice caused by mutation in *delta 2 glutamate receptor* gene. *Nature,* 1997, 388, 769-73.

CONTRIBUTORS

EDITOR/CONTRIBUTOR:

Robert Lalonde, Ph.D.
University of Rouen
Department of Psychology
ICONES Laboratory EA 4699, 76821
Mont-Saint-Aignan Cedex, France
TEL: +33 02 35 14 61 08
FAX: +33 02 35 14 63 49
Virtual International Authority File number: Lalonde, Robert 1954-
Virtual International Authority File: VIAF=102896257
E-mail: robert.lalonde@univ-rouen.fr
https://www.researchgate.net/profile/Robert_Lalonde/

CONTRIBUTORS:

Catherine Strazielle
Université de Lorraine
Laboratoire "Stress, Immunité, Pathogènes" and
Service de Microscopie Electronique, Faculté de Médecine,
9 avenue de la Forêt de Haye, 54500
Vandœuvre-les-Nancy, France
CHU de Nancy, 54500 Vandoeuvre-les-Nancy, France
E-mail: catherine.strazielle@univ-lorraine.fr
https://www.researchgate.net/profile/Catherine_Strazielle/

K. Fukuchi
University of Illinois College of Medicine at Peoria
Department of Cancer Biology and Pharmacology
P.O. Box 1649
Peoria Ilinois 61656, USA
E-mail: kfukuchi@uic.edu
https://www.researchgate.net/researcher/39437559_Ken-ichiro_Fukuchi/

INDEX

C

D

I

L

M

N

O

P

Q

R

S